Petrochemicals 2024

Topical Conference at the 2024 AIChE Spring Meeting and 20th Global Congress on Process Safety

New Orleans, Louisiana, USA
24-28 March 2024

ISBN: 979-8-3313-1715-7

Printed from e-media with permission by:

Curran Associates, Inc.
57 Morehouse Lane
Red Hook, NY 12571

CURRAN ASSOCIATES INC. proceedings .com

Some format issues inherent in the e-media version may also appear in this print version.

AIChE
120 Wall Street, FL 23
New York, NY 10005-4020

Phone: (800) 242-4363
Fax: (203) 775-5177

www.aiche.org

Additional copies of this publication are available from:

Curran Associates, Inc.
57 Morehouse Lane
Red Hook, NY 12571 USA
Phone: 845-758-0400
Fax: 845-758-2633
Email: curran@proceedings.com
Web: www.proceedings.com

TABLE OF CONTENTS

POSTER SESSION: FUELS AND PETROCHEMICALS DIVISION

59b Implementation of a Process Safety Management System and Inherently Safer Design Strategies for a Rehabilitated Bioethanol Facility in Trinidad and Tobago 1
Safiyyah Wahid

59c From Hydrogen Economy to Ccus and Beyond: Unleashing the Power of Digital Solutions in Sustainability 2
Ana Khanlari, Phd

59d Displacement Efficiency and Production Characteristics of in-House Developed Surfactants for Enhanced Oil Recovery 3
Syed Muhammad Shakil Hussain, Muhammad Shahzad Kamal, Shirish Patil, Ahmad Mahboob, Mobeen Murtaza, Mohammad Alotaibi, Mohanad Fahmi

59f Development of Nonionic Demulsifiers for Produced Water Treatment 4
Syed Muhammad Shakil Hussain, Muhammad Shahzad Kamal, Ahmad Adewunmi, Shirish Patil

59g Synthesis and Characterization of Magnetite Nps Decorated Onto ACF, Heat-Treated g-C_3N_4, Pure CS and OSCS Nps for the Efficient Removal of MB and CV Dyes from Aqueous Solution 5
Ashish Bhagat

59h Synthesis of Magnetic Nanoparticles Anchored on Carbon Nano-Matrix for the Efficient Adsorptive Removal of Crystal Violet from Aqueous Solution 6
Ashish Bhagat

PROCESS CONTROL AND OPTIMIZATION DEVELOPMENTS

75a Lessons Learned on Binary Distillation Column Control Scheme Design 7
James Beall IV

75b Feedforward Control Methods and Benefits 8
Greg McMillan

75c Successful Use of Digital Twin Technology and Advanced Control to Optimize Industrial Energy Systems 17
Ben Janvier, James Beall IV

DEVELOPMENTS IN PETROCHEMICALS

115a Exploring Propane Conversion Via Unconventinal Piston Reactor – a Modeling Study to Guide Process Development and Economics 18
Aya Abousrafa, Mary Katebah, Mamoun Al-Rawashdeh, Patrick Linke

115b Innovative Polymer-Surfactant Systems for Enhanced Oil Recovery in Challenging Reservoir Environments: Rheological Insights and Performance Assessment 20
Syed Muhammad Shakil Hussain, Shirish Patil, Muhammad Shahzad Kamal, Ahmad Mahboob, Mobeen Murtaza, Mohammad Alotaibi, Mohanad Fahmi

115c Phosphate intercalated layered double hydroxide as potential flame retardants for polystyrene 21
Farrukh Shehzad, Mamdouh A. Al-Harthi

REFINERY - PETROCHEMICAL INTEGRATION

136a Synergistic Optimization of Crude Oil Distillation Units and Heat Exchanger Networks: A Simultaneous Approach for Enhanced Energy Efficiency 22
Chenguang Zhu, Stephen Doyle, Zekun Yang, Julia Jimenez Romero, Nan Zhang

136b New Age Topping Plant for Regional Refining and Regional Power Generation 23
Randy Stubbs, Douglas Swain

136c Smart and Multifunctional Tubacoat Technology for CAPEX and OPEX Reduction, Decarbonization and Biofuels Production in Refining and Petrochemical Applications, ... 24
Sanjay Lodha

FEEDSTOCKS FROM RECYCLED MATERIALS

153a Low Temperature Routes to Catalytic Conversion of Polyethylene 25
Daniel Shantz

153b Tailoring HZSM-5 for Sustainable Waste Polypropylene Recycling: Enhancing Catalyst Stability and Selectivity in Light Olefin Production 29
Oguzhan Akin, Qing He, Yihan Wang, Parviz Yazdani, Robin John Varghese, Kevin Van Geem

153c Tandem Dechlorination and Hydrogenation of PVC to Waxes 42
Konstantinos Goulas, Scott Svadlenak

ADVANCES IN AMMONIA AS A FUEL/HYDROGEN CARRIER

184a CFD Modeling of Industrial Burners: Investigation of Hydrocarbon Fuel Vs. Carbon-Free Ammonia Based Fuel 43
Krishnadash S. Kshetrimayum, Minkyu Han, Younghee Lee

184b Process Design, Simulation and Techno-Economy Analysis of the Green Ammonia Production from Produced Water 44
Lucylitzy Bautista, Ana Gonzalez, Vahid Taghikhani, Mohammad Moosazadeh

184c Evaluating Alternative Energy Sources 45
Ian Sutton

PROCESS CONTROL AND OPTIMIZATION DEVELOPMENTS II

96a Bridging Planning, Scheduling, and Real-Time Operations: Expanding the Horizon of APC with Multiperiod Optimization 47
Filippo Trivella

96b Use Standard Material Balance Equations for Critical Level Control Applications. 48
Charles Herzog

96c Advancing Petrochemical Processes Inferential Analysis with Eigen's ML Transfomers 65
Erdin Guma

FEEDSTOCKS FROM RECYCLED MATERIALS II

172a Unconventional Feedstock Processing 66
Hsiang Yee Hoekstra, Eloy Flores III, Veshal Venkat

172b Influence of Feedstock Variability on Corrosion Dynamics in Municipal Wast-to-Biofuel Conversion Processes 67
Russ Davis

172c Life-Cycle Analysis of Recycling of Post-Use Plastic (PUP) with the Greet Model................................ 75
Pahola Thathiana Benavides, Ulises Gracida Alvarez, Troy Hawkins, Michael Wang

Author Index

59b Implementation of a Process Safety Management System and Inherently Safer Design Strategies for a Rehabilitated Bioethanol Facility in Trinidad and Tobago

Safiyyah Wahid

Over the last century, Trinidad & Tobago's (TT) economy has been primarily oil and gas based. Diversification of the local economy to improve foreign exchange earnings has been at the forefront of national development plans in recent decades. The introduction of the Caribbean Basin Initiative (CBI) in 2007 provided a valuable opportunity for diversification - the commissioning and operation of a 100 million US gallon/year bioethanol facility in Trinidad that was capable of denaturing 190 proof ethanol feedstock from Brazil for export to North America for use as a biofuel in vehicles. In 2012, the facility ceased its operations because of a discontinuation of its feedstock supply and contributory global market factors. Based on the vibrant sugar cane industry that TT once enjoyed, a business model was proposed which involved using local feedstock to continue and sustain the facility's operations, which would have also resulted in the introduction of bioethanol fuel into the local transportation sector with the benefits of reducing gasoline imports and promoting cleaner vehicular emissions. However, this strategy was not viable largely due to the diesel and gasoline fuel subsidy enjoyed by citizens at the time.

With the recent economic downturn due to the Covid-19 pandemic, and the emphasis the Government of TT has placed on the transition towards a diversified economy in its Vision 2030 National Development Strategy, the revitalization of the sugar cane industry in TT cannot come at a better time. The government's removal of the local fuel subsidy is also a catalyst for national discussions to begin on the use of bioethanol fuel and subsequently, the rehabilitation of this facility. Should rehabilitative works prove to be economically viable, a rigid Process Safety Management (PSM) system with the application of Inherently Safer Design (ISD) strategies must govern the modifications to this aged plant, as well as the production operations to ensure compliance with modern industry safety standards. The objective of this study is to therefore apply the 14 elements of PSM to this rehabilitated and modified bioethanol facility, and employ ISD strategies for risk reduction and incident prevention. An added benefit of this study is that this PSM model can be applied to other aged, local processing facilities which currently do not follow a PSM standard, which is one step closer to improving the overall safety culture in industrial organizations in TT.

Keywords: *Bioethanol, Process Safety Management, Inherently Safer Design*

59c From Hydrogen Economy to Ccus and Beyond: Unleashing the Power of Digital Solutions in Sustainability

Ana Khanlari, Phd

With net zero targets and circular economy initiatives, the energy and chemical industry is navigating the energy transition landscape. The industry is depending on digital tools to accelerate the pace of innovation and scale up of new solutions. Whether these solutions are minimizing energy use and emissions, incorporating biobased or recycled feedstocks, diversifying energy sources or capturing CO2, digital solutions provide a groundwork for economic feasibility as well as design, scale up and operation of the new or existing assets. This presentation will provide a holistic view of the different sustainability pathways where digital solutions play a major role.

59d Displacement Efficiency and Production Characteristics of in-House Developed Surfactants for Enhanced Oil Recovery

Syed Muhammad Shakil Hussain, Muhammad Shahzad Kamal, Shirish Patil, Ahmad Mahboob, Mobeen Murtaza, Mohammad Alotaibi, Mohanad Fahmi

Enhanced oil recovery (EOR) is a technique in which chemicals are injected into the formation to recover over 60% of the remaining oil after waterflooding. Surfactants are the main components used for chemical-enhanced oil recovery (cEOR). However, surfactant adsorption as well as precipitation due to elevated temperature and salinity reservoir environments are the significant challenges the industry facing today. The current research intends to systematically experiment with the influence of elevated temperature and high salinity on surfactant adsorption and provides a guideline for controlling their retention and precipitation due to harsh conditions.

In the first step, the zwitterionic surfactants were prepared using green or no solvents and characterized by NMR, FTIR, and ESI-MS tests to verify the structure of the surfactant and validate the synthetic method. The heat tolerance capacity of the surfactant was measured by Thermal Gravimetric Analysis (TGA) as well as the aging technique. The surface/interfacial tension was studied using a force tensiometer and spinning drop method, respectively. The adsorption tests were conducted on calcite, dolomite, and quartz. The morphology of the rocks was characterized by X-ray diffraction (XRD) and Scanning Electron Microscopy (SEM) methods.

The in-house prepared surfactants displayed high solubility in high and low-salinity water without any precipitation over a long period. The TGA curves revealed a sharp decline above 250 ^{o}C which was greater than the reservoir temperature ($\geq$90°C). In addition, the aging technique showed no phase separation or milkiness of the aqueous solution of surfactants after keeping the samples in the oven for up to three months at 90 ^{o}C. The surface/interfacial tension of the prepared surfactants was similar to or larger than the commercial surfactants. Experimental data indicates that the surfactant adsorption varies by varying the clay minerals due to intercalation of surfactant on the carbonates.

The results revealed that the generated surfactants exhibited high heat stability, clear aqueous solutions, low surface/interfacial tension, and less adsorption onto carbonates. All these advances are important in cEOR that will help to develop a suitable material to enhance production with reduced cost and time.

59f Development of Nonionic Demulsifiers for Produced Water Treatment

Syed Muhammad Shakil Hussain, Muhammad Shahzad Kamal, Ahmad Adewunmi, Shirish Patil

In the realm of crude oil exploration in oilfields, managing water production poses a substantial challenge due to the prevalence of heavy emulsions, brought about by natural emulsifiers present in crude oil. The breaking of such emulsion is important before transporting the crude oil to refineries. This study investigate the emulsion-breaking capabilities of a freshly developed nonionic demulsifier. Its purpose is to efficiently separate oil and water by disrupting emulsifying agents. Demulsification efficiency (DE%) was assessed using the bottle test technique at both room and moderate temperatures, while rheological measurements were also conducted on emulsions before and after demulsification to validate the bottle test outcomes. The introduction of the newly developed nonionic demulsifier at various concentrations effectively broke the emulsion, resulting in water-oil separation. Remarkably, these cost-effective chemicals achieved demulsification efficiencies exceeding 95%. The rheological measurements demonstrated the migration of this nonionic demulsifier to the water-oil interface, where it interacted with and destabilized natural emulsifiers responsible for crude oil emulsion stability. Importantly, this research introduces a nonionic chemical demulsifier capable of treating emulsions produced in oilfields, efficiently separating oil and water into distinct phases.

59g Synthesis and Characterization of Magnetite Nps Decorated Onto ACF, Heat-Treated g-C_3N_4, Pure CS and OSCS Nps for the Efficient Removal of MB and CV Dyes from Aqueous Solution

Ashish Bhagat

Industrial activities lead to the release of toxic dye compounds into the water bodies on a regular basis. Toxic dyes are known to create health hazards to the mankind and the aquatic ecosystems. This current study describes the novel synthesis of magnetite (Fe_3O_4) nanoparticles (NPs) decorated onto various carbonaceous materials, viz., pure activated carbon fiber (ACF), heat-treated graphitic carbon nitride (g-C_3N_4-HT), candle soot (CS) particles, and onion shaped candle soot (OSCS) particles using coprecipitation technique. The physicochemical properties of the synthesized materials are extensively investigated using various analytical and spectroscopic techniques such as SEM, TEM, VSM, EDAX, FTIR, XRD, XPS, and BET. Characterization data confirm that magnetite NPs are successfully incorporated onto pure ACF, g-C_3N_4-HT, CS, and OSCS surfaces. The synthesized materials are tested for their efficacy towards the removal of two industrially relevant dyes, namely, methylene blue (MB) and crystal violet (CV) dyes from aqueous solutions. The experimental data show that all four materials are capable of removing both the dyes. The adsorption performance of all four materials satisfy the following trend:

Fe_3O_4-ACF > Fe_3O_4-OSCS > Fe_3O_4-CS > Fe_3O_4/g-C_3N_4-HT

The batch adsorption kinetic data for all four adsorbents agreed well with the pseudo-second order kinetic model. Besides, Langmuir model is the best fitted isotherm model for all the adsorbents. Owing to the magnetic property, the adsorbents are efficiently separated from the aqueous solution with the help of an external magnet. This obviates the requirement of time consuming centrifugation and filtration steps. Additionally, the spent adsorbents are further characterized using FTIR, XRD, and SEM analyses. The characterization data reveal that no structural deterioration /chemical transformation of adsorbents occur post adsorption studies. With adequate optimization, the materials may be utilized to remove recalcitrant pollutants from water.

59h Synthesis of Magnetic Nanoparticles Anchored on Carbon Nano-Matrix for the Efficient Adsorptive Removal of Crystal Violet from Aqueous Solution

Ashish Bhagat

In this study, we present a substrate-free synthesis method for generating Cu-carbon nanofibers (Cu-CNFs) using Cu-tartrate. The optimization of the reduction temperature of Cu-tartrate through temperature programmed reduction (TPR) analysis is identified as a critical factor for the successful growth of Cu-CNFs. Additionally, magnetite (Fe3O4) nanoparticles (NPs) are incorporated onto Cu-CNFs using the wet-impregnation technique. Various analytical and spectroscopic techniques are employed to evaluate the physico-chemical properties of the resulting materials.

The material is subjected to testing for its adsorption efficacy towards crystal violet dye. A notable adsorption capacity (260 ± 5 mg/g at a pH value of 11, as measured by the Langmuir model) is observed, attributed to the combined effects of a negative surface charge, hydrophilic characteristics, and a high surface area of the material. In a dual role, Fe3O4 NPs facilitate the adsorption of crystal violet onto the adsorbent and efficiently separate the adsorbent from the aqueous mixture using an external magnet. The synthesized material exhibits potential for the adsorptive removal of other cationic dyes from wastewater.

75a Lessons Learned on Binary Distillation Column Control Scheme Design

James Beall IV

There are over 40,000 distillation towers in operation in the United States. Distillation is a low thermal efficiency unit operation that currently consumes 4.8 quadrillion BTUs of energy—40% of the processing energy used in refining and continuous chemical processes. The multiphase gas/liquid flow patterns in distillation columns are complex, making them difficult to predict and control.

The control scheme for the column has 3 objectives as follows.

1. Produce product streams that meet their respective quality specifications
2. Perform the component separation at the lowest energy consumption
3. Operate with the least amount of operator intervention

For the basic binary distillation column with a constant feed, there are 24 combinations of the regulatory control schemes. Which do you choose to achieve the objectives of the column listed above? If you implement Advanced Multivariable Control, which underlying regulatory control scheme and manipulated variables are the best?

This presentation will explore these topics and provide some insight into what has worked well and what has not! Plant examples will be used to demonstrate the concepts.

SPRING24
+20TH GCPS
A Joint AIChE and CCPS Meeting

Feedforward Control Methods and Benefits

Gregory K McMillan
Retired Principal Senior Software Engineer
11404 Dodge Cattle Dr
Austin, Tx 78717
gkmcmi@austin.rr.com

Prepared for Presentation at
American Institute of Chemical Engineers
2024 Spring Meeting and 20th Global Congress on Process Safety
New Orleans, LA
March 24-28, 2024

Feedforward Control Methods and Benefits

Gregory K McMillan
Retired Principal Senior Software Engineer
11404 Dodge Cattle Dr
Austin, Tx 78717
gkmcmi@austin.rr.com

Keywords: load disturbances, ratio control, reactor control, steam header control

Abstract

Guidelines and test results are provided on feedforward control dynamic compensation and configuration to improve load and setpoint responses. The prevalent confusion on feedforward multiplier and summer use is explained and resolved. The methods and benefits of upgrading to ratio control including online adaptation are detailed and the extension to plantwide control is noted. The strategies for preemptive action needed in procedure automation and in the open loop backups critical for preventing compressor surge and Resource Conservation and Recovery Act (RCRA) pH violations are addressed.

1 Introduction

Advanced regulatory control seeks to incorporate knowledge of process dynamics, disturbances, constraints, and objectives to increase process efficiency and capacity. The PID power and flexibility enables an incredible spectrum of creative opportunities to achieve these goals.

The most common advanced control technique is feedforward control that is designed to preemptively correct for a disturbance. In industrial applications, feedforward control is far from perfect so feedback control is essential. Most of the problems with feedforward control stem from incorrect and missing dynamic compensation or poor measurement and valve resolution or rangeability. Separate open loop tests should be conducted where the disturbance variable and manipulated variables are stepped. If a time constant plus dead time model of both responses is adequate, the dynamic compensation simplifies to a feedforward dead time and lead-lag block.

2 Fundamentals

Feedforward control is the most frequently used advanced regulatory control technique. The most common feedforward signal is a feed flow. The result is flow ratio control with feedback correction by means of a feedforward summer. For better operator involvement, the operator must be able to change the desired ratio and see the actual ratio. This operator interface is particularly important for running on just ratio control (no feedback correction) for startup, abnormal operation, and analyzer problems. While the literature leads one to believe a feedforward multiplier is better for changes in ratio, feedforward correction multiplier signal scaling is problematic and the gain nonlinearity introduced into liquid systems with back mixing is undesirable. In well-mixed volumes, effect on PID tuning of the change in process gain with production rate is compensated by the change in the primary time constant (residence time) with production rate. Finally, most of the error in measurements is an offset error that can be corrected by a bias. The fact that MPC and neural networks use a bias correction is a clue to fundamental advantages of a feedforward summer over a feedforward multiplier.

The feedforward signal needs dynamic compensation so that the feedforward correction arrives at the same point at the same time in the process as the disturbance as depicted in Figure 1. Note that in the process industry, disturbances occur in stream flows, compositions, and physical properties and are termed load upsets.

The feedforward dead time is the disturbance variable dead time minus the manipulated variable dead time. If this value is negative, dynamic compensation of dead time is not used and the feedforward gain needs to be reduced because the feedforward correction will be late and some feedback action will already have occurred.

The feedforward lag time is set equal to the disturbance variable time constant. The lead time is then set equal to the manipulated variable time constant. The ratio of manipulated variable to disturbance variable time constant is the lead-lag factor. If this factor is greater than one, there is a kick in the feedforward to compensate for the excessive slowness in the manipulated variable. If open loop tests cannot be done or higher order dynamics (e.g., secondary lag) alter the initial response or an integrating or runaway response occur, then an addition of lead or lag time to make the feedforward response faster or slower may be the simplest solution. Test results will show how the closed loop response to a disturbance can be used to decide what term to adjust and in what direction.

If the feedforward arrives too soon, an inverse response is created that is particularly disruptive. In this case, the feedforward can do more harm than good. Since timing cannot be perfect, a feedforward signal that is a little late is better than one that is early especially considering PID controllers are typically tuned with a gain that is far below the maximum. Thus, the better solution is to make sure the feedforward signal can handle major disturbances with the largest dead time and slowest time constant by being liberal with the use of dead time and lag time in the manipulated variable path. This is not to say there is no problem with a feedforward dead time or lag that is excessive. If the feedforward correction arrives late by more than half the ultimate period, the feedforward correction can get in phase with the feedback correction, creating undershoot. If the feedforward correction is more than twice the ultimate period, a second upset in the opposite direction of the original upset is created. The amplitude of undershoot or a secondary upset can approach the amplitude of the original disturbance depending upon the feedforward gain. Knowledge of how time constants and dead times change with production rate should be used to schedule the settings in the feedforward dynamic compensation to avoid the feedforward signal arriving early or too late.

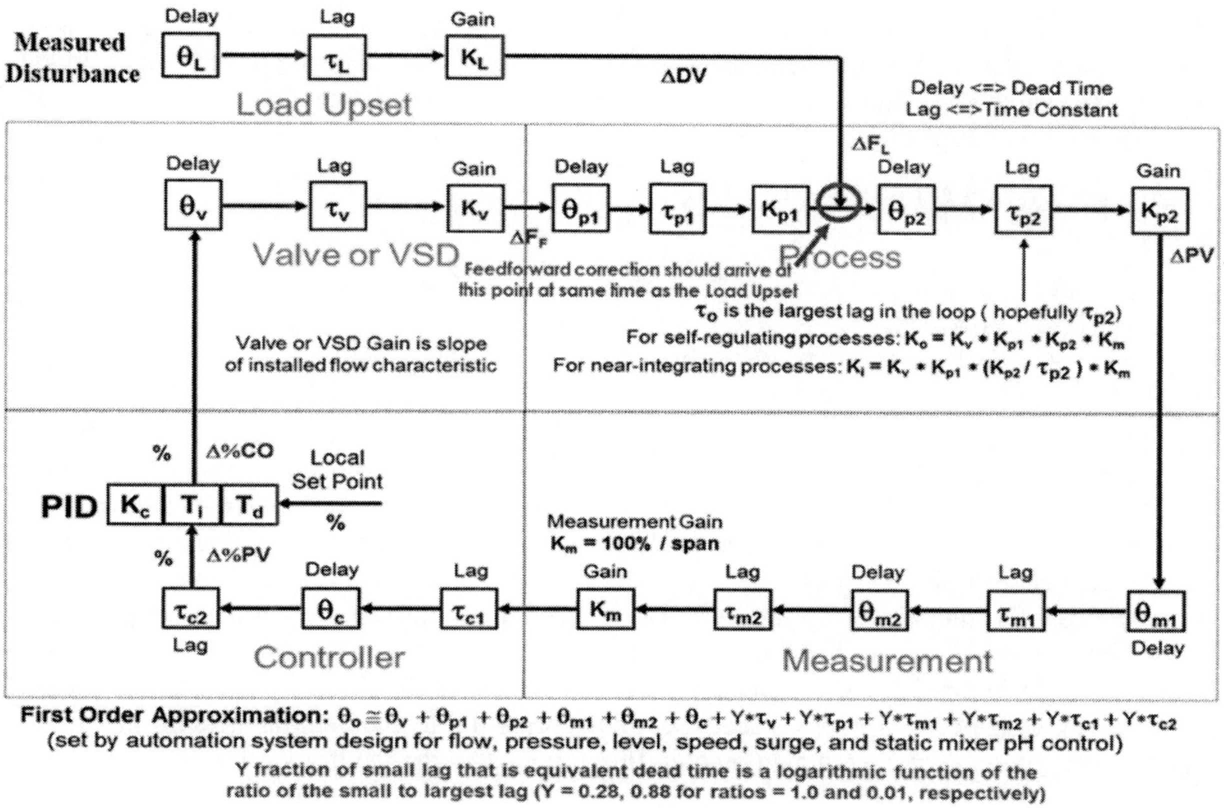

Figure 1. Loop and Load Disturbance Block Diagram

For the general case of feedforward, where the controller output scale is set as CO_{max} and CO_{min} and the feedforward scale is set as DV_{max} and DV_{min}, the feedforward gain (K_{ff}) is seen in the following equations and can be identified by a tuner setup to identify the $\Delta\%PV$ for a $\Delta\%DV$ with the feedforward scale set.

$$K_{FF} = \frac{\frac{\Delta\%PV}{\Delta\%DV}}{\frac{\Delta\%PV}{\Delta\%CO}} = \frac{\Delta\%CO}{\Delta\%DV}$$

$$K_o = \frac{\Delta\%PV}{\Delta\%CO}$$

$$\Delta\%CO = \frac{100\%}{CO_{\max} - CO_{\max}} * \Delta CO$$

$$\Delta\%DV = \frac{100\%}{DV_{\max} - DV_{\min}} * \Delta DV$$

For a feedforward that is essentially flow ratio control, the K_{ff} simplifies to be simply the mass flow ratio R. if the feedforward scale is set equal to the valve flow capacity or secondary flow loop span. If the valve is nonlinear, a signal characterizer on the PID output should be used to provide the linear flow relationship.

$$\Delta F_F = K_V * K_Z * R * \Delta F_L$$

$$\Delta F_F = \frac{F_{V\,max}}{100\%} * \frac{100\%}{F_{V\,max}} * R * \Delta F_L$$

$$\Delta F_F = R * \Delta F_L$$

$$K_{FF} = R$$

The most common feedforward signal is flow because of production rate changes and because nearly all fast and large disturbances originate as flow rather than composition or temperature disturbances. To attain the process flows on the process flow diagram (PFD), the manipulated flow is ratioed to a wild or leader flow. In most cases, the feedforward is the primary feed flow. A ratio station is used to provide the operator with the ability to set the desired ratio and see the actual ratio. The process PID provides a feedback bias correction to the manipulated flow. The actual ratio displayed is the corrected manipulated flow divided by the feedforward flow.

Setpoint feedforward does not require dynamic compensation. Setpoint feedforward only offers a significant advantage if the step change in PID output from the PID modes for a setpoint change is less than 50 percent of the final change needed as a step change in PID output. This scenario can be caused by a structure of proportional on process variable (PV) (I on error and PD on PV), the product of the beta setpoint weight factor and PID gain less than 0.5 for a two degrees of freedom (2DOF) structure, or the product of the setpoint lead to lag ratio and PID gain less than 0.5 for a structure of proportional on error (e.g., PI on error and D on PV). These values are offered as a guideline only. Specific application conditions need to be considered.

The other major considerations to be addressed are disturbance measurement noise, error, and rangeability. Noise can be amplified by a feedforward lead time greater than the lag time. If the rangeability of a flow measurement is insufficient, a flow calculated from valve position should be used for low flows. A flow setpoint instead of a flow measurement for the flow feedforward will eliminate noise and provide coordination of flow changes for a change in production rate.

Intelligent output action via output tracking and the remote output mode enable the PID output to be scheduled to deal with abnormal situations. This function is called an open loop backup because the feedback control is temporarily suspended to guarantee a fast correction. Control is returned to the PID feedback control when a protective state is assured. An open loop back up has been extensively used to prevent surge and environmental violations.

The PID output can be set at a limit to achieve the fastest possible setpoint response without upsetting other loops by setpoint rate limiting of the manipulated flow. The PID output is then set to a final resting value (captured from last batch or startup) and control is returned to the PID when a PV value one dead time or more into the future is projected to reach the setpoint. This full throttle response can reduce batch cycle and startup times.

3 Opportunities

The opportunities all stem from a preemptive correction of a disturbance. The feedforward correction needs to arrive at the same time at the same point in the process as the disturbance with an equal magnitude and opposite sign.

1. A feedforward timing and signal accuracy of 10 percent offers a 10:1 improvement in the peak and integrated error for a load disturbance (e.g., feed disturbance).

2. Plantwide feedforward is possible where the plant moves in concert to production rate and product grade changes and to minimize inventories.

3. PID tuning can be less aggressive and consequently less oscillatory and more robust since most of the work is done by feedforward.

4, Startup can be on ratio control until equipment reaches operating point to start PID control.

4 Watch-outs

Most of the watch outs are associated with feedforward measurement or valve problems, feedforward timing mismatch, or unmeasured disturbances. All changes must be thoroughly tested.

1. Limit cycles in the feedforward signal from valve stick-slip particularly near the closed position which leads to poor rangeability.

2. Low signal to noise ratio in the feedforward signal at low velocities which leads to poor rangeability.

3. Manipulated variable dead time that is more than twice the disturbance dead time leading to severe undershoot or a secondary disturbance from feedforward correction arriving late by more than half the ultimate period.

4. An unmeasured disturbance (e.g., change in feed composition) coincident with a change in feed flow that causes an upset that is larger in the opposite direction. Here an intelligent feedforward can recognize that a feedforward correction based on the flow change would cause a PV change in the same direction of current excursion from a different disturbance and should be moderated.

5 Recommendations

1. Use feedforward control to give preemptive action for large and fast disturbances.

2. Use ratio and bias and gain stations to provide flow ratio control with feedback bias correction and the actual flow ratio viewable and desired flow ratio accessible by operations. Enable startup on ratio control without feedback correction until the process (e.g., distillation column) reaches operating condition.

3. Provide dynamic compensation of feedforward signals by dead time and lead-lag blocks to give the right feedforward timing for disturbances.

4. If open loop tests cannot be done or the process response is complex (e.g., high order or recycle effects), integrating, or runaway, the dead time and lead-lag times can be adjusted based on oscillation patterns observed.

5. Ensure the feedforward correction does not arrive too soon causing an inverse response that can result in the feedforward doing more harm than good.

6. If the feedback correction will unavoidably arrive late, decrease the feedforward gain to prevent a secondary oscillation and undershoot that result from feedback control having already done some of the correction.

7. Use intelligent output action via output tracking or remote output to schedule output changes for an open loop backup to protect equipment, minimize environmental violations, or prevent Safety Instrumented Systems (SIS) trips or to enable a full throttle setpoint response (bang-bang control) to minimize startup and batch cycle times. In an open-loop backup, the PID is momentarily put in output tracking with a future value block employed to provide a rapid, proactive, predictive correction that has been particularly shown to prevent an RCRA *p*H violation or compressor surge with a smooth transition back to feedback control. The future value block computes a future value by computing an appreciable change as the output minus the input to a deadtime block with a deadtime setting to optimize signal to noise ratio and then multiplying the resulting rate of change by the total loop deadtime and simply adding the result to the current deadtime block input to get an intelligent future value.

8, Consider a simple integral-only controller to gradually adapt a feedforward gain or ratio setpoint seeking the minimize the bias correction needed by the process feedback controller. For a feedforward multiplier, the adaptive controller optimizes the feedforward or ratio station bias to minimize the feedforward gain or ratio station bias correction needed by the feedback controller.

6 Test Results

Test results shown in the presentation were generated using a DeltaV Digital Twin (virtual plant) with the ability to set the process type and dynamics, automation system dynamics, PID options (structure and enhanced PID), PID execution time, setpoint lead-lag, tuning method, and a step change in lower PID load (LF) or upper PID setpoint (SP). Table 1 summarizes the test conditions.

The disturbance delay was 20 seconds for all the tests. This is the delay between the measurement of the load disturbance and the arrival at the same point in the process as the manipulated flow from the control valve. For these tests, the disturbance entered the process downstream of the process delay and secondary time constant but upstream of the primary process time constant. Consequently, the delay between a change in feedforward measurement and the arrival of the control valve flow is the sum of the feedforward signal delay and the process dead time. In this case the feedforward delay should be equal to the disturbance delay less the process dead time.

The disturbance lag was 10 seconds for all the tests. The 86 percent response time of the control valve was about two seconds. The manipulated flow from the control valve passed through a secondary process time constant of one second before arriving at the same point as the disturbance that is upstream of the primary process time constant. In this case the feedforward lag should be set equal to the disturbance lag (10 seconds) and the feedforward lead should be set equal to the control valve time constant (half the 86 percent valve response time) plus the secondary time constant for a total of two seconds.

Balanced (moderate self-regulating_ and lag dominant self-regulating (near-integrating) processes are used in the tests because these are the more prevalent types of processes in continuous plants and to show the effect of the difference between the process and feedforward time constant (lag). The results in Tests 1a, b and Tests 2.a, b show that a feedforward signal that arrives too soon or too late due to a feedforward dead time or lag mismatch of 10 seconds will as expected cause an inverse response and an undershoot, respectively. The effect is worse for a moderate self-regulating process. For very fast processes and dead time dominant processes, the consequences of feedforward timing error are even more severe.

The results in Tests 3a, b show that the consequences of a feedforward gain being too large are much more disruptive than a gain too low. If there is no initial response in the right direction and the response in the wrong direction is obvious, the feedforward gain is too large.

Table 1. Test conditions

Tests	Process type	Open loop gain	Delay (sec)	Lag (sec)	Change	PID Tuning
1a	Moderate self-reg.	1 dimensionless	10	20	$\Delta F_L = 20\%$	Feedforward delay
1b	Near-integrating	1 dimensionless	10	100	$\Delta F_L = 20\%$	Feedforward delay
2a	Moderate self-reg.	1 dimensionless	10	20	$\Delta F_L = 20\%$	Feedforward lag
2b	Near-integrating	1 dimensionless	10	100	$\Delta F_L = 20\%$	Feedforward lag
3a	Moderate self-reg.	1 dimensionless	10	20	$\Delta F_L = 20\%$	Feedforward gain
3b	Near-integrating	1 dimensionless	10	100	$\Delta F_L = 20\%$	Feedforward gain

See the presentation for much more details and examples including the important applications for steam header control, reactor control, and compressor surge control.

7 Resources

[1] McMillan, G.K. and Vegas, P.H., *Process/Industrial Instruments and Controls Handbook Sixth Edition*, New York, McGraw-Hill Professional, 2019, https://www.mhprofessional.com/9781260117974-usa-process-industrial-instruments-and-controls-handbook-sixth-edition [mhprofessional.com]

[2] McMillan, G.K., *Tuning and Control Loop Performance Fourth Edition*, New York, Momentum Press, 2015, (free pdf copy by emailing author).

[3] McMillan, G.K., "Foray into feedforward control", *Control*, January, 2023, https://www.controlglobal.com/control/loop-control/article/21460548/control-talk-foray-into-feedforward-control

[4] McMillan, G.K., "Realizing ratio control", *Control*, March, 2022, https://www.controlglobal.com/measure/flow/article/11288360/realizing-ratio-control

[5] McMillan, G.K. and Vegas, P.H., "Best practices for PID", *Control*, December, 2018, https://www.controlglobal.com/control/loop-control/article/11304417/best-practices-for-pid

[6] McMillan, G.K., "Feedforward and Ratio Control", ISA Mentor Program Webinar, 2017, https://www.youtube.com/watch?v=7kt_jFlnlDg&list=PLC3fVVaSjwYFSYiSSUJuIUL4vHMiQ6hT1&index=12

[7] McMillan, G.K., **"**Feedforward Control Enables Flexible, Sustainable Manufacturing**"**, *InTech***,** March 2011, (free pdf copy by emailing author)

[8] McMillan, G.K., Control Talk Column https://www.controlglobal.com/control-talk-column

[9] McMillan, G.K., Ask the Automation Pros https://blog.isa.org/tag/ask-the-automation-pros

75c Successful Use of Digital Twin Technology and Advanced Control to Optimize Industrial Energy Systems

Ben Janvier, James Beall IV

Modern technology has provided chemical and petrochemical plants with the option of using Digital Twins in their day-to-day operations. Indeed, Digital Twins can nearly perfectly replicate plant dynamics in a virtual environment when first-principle models are used during the design phase. And, once developed, they can be incorporated into several of the plant's strategic goals: safety, reliability, training, profitability and carbon footprint reduction.

This paper will present a few examples of the use of Digital Twins on industrial energy systems:

- Increase reliability and safety by testing and commissioning critical assets (BMS, island shedding logic, Advanced Controls, etc.)
- Implementation of Advanced Controls to stabilize and optimize operation
- Improve a facility's carbon footprint by implementing a Dynamic Energy Management System
- Improve operator training using a Digital Twin.

115a Exploring Propane Conversion Via Unconventinal Piston Reactor – a Modeling Study to Guide Process Development and Economics

Aya Abousrafa, Mary Katebah, Mamoun Al-Rawashdeh, Patrick Linke

The commodity chemical industry is a major consumer of energy and is responsible for producing around 25% of global greenhouse gas emissions [1, 2]. Hence, decarbonization of the chemical industry is an imperative step towards achieving a sustainable and carbon-neutral energy system. With the growing prominence and cost-effectiveness of renewable energy sources, the paradigm shift toward electrification via utilizing novel reactor systems that exploit renewable electricity to drive chemical processes appears to be the most viable strategy for reducing the carbon footprint of chemicals[3]. The ability to convert electrical energy into chemical energy offers a means to store electricity and overcome the intermittency of renewable sources such as solar and wind.

The piston reactor is an innovative concept that converts electrical energy into mechanical work and, subsequently, into chemical energy storage molecules[4, 5]. It operates much like a conventional combustion engine, cyclically compressing and expanding a feed gas within very short timeframes, typically in the millisecond range. This compression generates high temperatures, reaching up to 1500 K, and pressures of several hundred bars, which can initiate reactions to produce common chemicals. The rapid gas expansion leads to quick cooling, helping the reactor maintain a desired non-equilibrium state and preventing secondary reactions of metastable species that could yield undesired by-products[6]. In addition to its capabilities, piston reactors are known for their simplicity, compactness, safety in operation, and ability to promptly adapt to variations in feed conditions[7].

The piston reactor can produce a range of chemical products through different reaction pathways. Therefore, it is crucial to identify promising products and their associated reactions from a multitude of emerging possibilities through early-stage evaluation. This assessment is vital for the early screening of promising directions, helping avoid extensive and time-consuming experimental work. The experimental landscape for the piston reactor is vast, involving various operational conditions, making it challenging to pinpoint favorable operating regions and the best pathways. Additionally, evaluating the feasibility of a reaction pathway based solely on experimental metrics like conversion, yield, or selectivity is inadequate. A more comprehensive analysis, considering economic and environmental factors like separation, raw material, product costs, and emissions, is essential. However, a significant challenge in early-stage assessment is the lack of complete data, particularly related to mass and energy balances necessary for a proper process evaluation. Therefore, there is an urgent need for a tool that can quickly and informatively assess and chart the operational space of the piston reactor for specific chemistry, facilitating key decision-making during the laboratory phase.

The current study presents an approach for a quick preliminary assessment of the piston reactor technology to investigate the opportunities and constraints of this reactor concept. For this purpose, a zero-dimensional single-zone timed-dependent thermodynamic model of a single-cylinder piston reactor is implemented. The kinetics are predicted using a detailed gas-phase mechanistic model. Propane is selected as a representative case study because it yields a diverse array of industrially relevant products including but not limited to hydrogen, ethylene, and propylene. The model is used to analyze the solution space which is then evaluated based on economic metrics to identify preferred operating regions. The exploration methodology follows a systematic, model-driven innovation approach to tackle the challenge of assessing opportunities arising from the intermediate mixtures produced during the reactor. Additionally, this work not only aims to develop a novel equipment concept but also to propose a comprehensive packaged solution capable of converting raw materials into value-added products.

In Figure 1, a case study is presented where the piston reactor operates with a propane/argon mixture to explore the potential for endothermic propane pyrolysis. The introduction of argon as a diluent in the feed mixture effectively reduces the heat capacity of the incoming gas, leading to higher temperatures during the compression phase, thereby enhancing conversion. This process generates a mixture consisting of hydrogen and various hydrocarbons, including methane, ethane, and ethylene within the analyzed solution space. Figure 1b provides a preliminary assessment of the added value of these products, considering their price per unit of amount of material fed (propane in this case). This evaluation can be extended to other cases involving different feed combinations or operating conditions, allowing for a comparative analysis of their added value. Additional economic indicators, such as separation costs and feed costs, can be incorporated into the evaluation to identify economically viable regions for further exploration.

115b Innovative Polymer-Surfactant Systems for Enhanced Oil Recovery in Challenging Reservoir Environments: Rheological Insights and Performance Assessment

Syed Muhammad Shakil Hussain, Shirish Patil, Muhammad Shahzad Kamal, Ahmad Mahboob, Mobeen Murtaza, Mohammad Alotaibi, Mohanad Fahmi

Oil extraction from reservoirs primarily relies on natural pressure or water injection, recovering just a fraction of the available oil (10% to 25%). Enhanced oil recovery (EOR) techniques, involving the injection of various chemicals such as surfactants, polymers, and alkalis, are utilized to increase hydrocarbon recovery. Polymers are used to increase water viscosity, thereby improving the oil/water mobility ratio, while surfactants modify the interfacial tension between water and oil. Often, a combination of surfactants and polymers is injected for improved mobility and reduced interfacial tension. However, creating a robust polymer-surfactant system capable of withstanding harsh reservoir conditions presents a significant challenge.

In our research, we synthesized hydroxybetaine zwitterionic surfactants and employed acrylamide-based copolymers with thermal stability and salt tolerance. We conducted a comprehensive rheological analysis of the surfactant-copolymer blend using an Anton Paar rheometer. Our findings indicated the compatibility of these polymeric surfactants and acrylamide-based copolymers at elevated temperatures, with no observable precipitation. The surfactant effectively reduced the interfacial tension of the oil-water interface. Furthermore, the rheological properties of the polymer-surfactant system exhibited a stable viscosity profile over an extended period, along with consistent elastic modulus values.

In deionized water, the addition of the surfactant resulted in reduced polymer viscosity due to charge screening. Remarkably, in seawater, the surfactant did not impact the viscosity of the polymer. Notably, oil recovery experiments demonstrated a substantial improvement, with over a 15% increase in oil recovery achieved through the injection of our novel surfactant-polymer system. Our study paves the way for the development of more effective EOR methods in challenging reservoir environments.

115c Phosphate intercalated layered double hydroxide as potential flame retardants for polystyrene

Farrukh Shehzad, Mamdouh A. Al-Harthi

The need to produce zero-energy buildings is becoming more popular in the context of energy resource conservation and demand reduction. Recently, the growing surge in the application of lightweight foamed concrete in the building sector, highlights an underlying potential in this area. Styrofoam beads (foamed polystyrene beads) are the commonly used material for such applications because of their low cost, availability, and thermal insulation property. However, polystyrene is highly flammable, and thus poses a safety risk. In this study , polystyrene was compounded with LDH (based on Al^{+}, Mg^{+} & Zn^{+}) and phosphate intercalated LDH to improve its flame retardance. LDH were synthesized by co-precipitation method and phosphate intercalation was achieved by ion-exchange method. The resulting composite were tested for its flame resistance by limiting oxygen index (LOI) and thermal stability. The limiting oxygen index was increased by nearly 5 % along with reduced flame spread, smoke generation and dripping. Overall, the study highlights the potential of modified LDH as a flame retardant for polystyrene and replacing the currently used halogenated flame retardants which are not environmentally friendly.

136a Synergistic Optimization of Crude Oil Distillation Units and Heat Exchanger Networks: A Simultaneous Approach for Enhanced Energy Efficiency

Chenguang Zhu, Stephen Doyle, Zekun Yang, Julia Jimenez Romero, Nan Zhang

Crude oil distillation, a substantial energy-consuming process, necessitates innovative approaches to enhance efficiency. This work introduces a simultaneous optimization strategy for both distillation units and Heat Exchanger Networks (HEN), differentiating itself by leveraging synergies within the system. Traditional methods sequentially optimize Crude Distillation Units (CDUs) and Vacuum Distillation Units (VDUs) before addressing the HEN, missing potential synergistic benefits. This technology, employs Artificial Neural Networks (ANNs) for distillation units, offering robustness, simplicity, and faster convergence compared to traditional models.

The simultaneous optimization considers variables like pump-around and stripping steam flow rates, coil outlet temperature, product flow rates, and HEN stream bypasses. By collectively optimizing these variables, The tool achieves superior outcomes in reducing energy costs, increasing product yields, and overall profitability. Notably, the technology accommodates various objectives, including minimizing energy consumption, maximizing profits, and adhering to practical constraints such as equipment capacities and product quality specifications.

Key distinguishing features include the simultaneous optimization approach and the use of ANNs for distillation models. Additionally, this technology considers practical constraints in HEN, offering a comprehensive solution for the complex challenges of crude oil distillation. This approach, represents a significant step towards advancing energy efficiency and economic viability in the crude oil distillation process, making it a potent tool for refinery optimization.

136b New Age Topping Plant for Regional Refining and Regional Power Generation

Randy Stubbs, Douglas Swain

History of Topping Plants:

The US military began operating topping plant refineries in remote locations during World War II to provide fuel for the war effort. In 1944 General MacArthur requested the US Army develop the Klamano oil fields in Papua New Guinea to provide basic atmospheric refining to produce diesel and aviation fuel for the continued Allied advance across the Pacific.

Since that time, topping plant refineries have evolved from crude 1st Generation open flame distillation of crude oil to 2nd Generation - Modular Refineries using fired process heaters and smaller footprints, and finally to the 3rd Generation – Turbo Refinery that utilize a gas turbine generator at its core and recovers waste heat from the turbine exhaust, through CHP to distil crude oil while simultaneously producing robust, excess electric power.

The heart of the Turbo Refinery is a simple atmospheric distillation column (ADU) for crude oil, which is exactly the same as a topping plant refinery, that has been in service since WWII. As such this technology has been in use for nearly 80 years in hundreds of small and large refineries worldwide.

This 3rd Generation Turbo Refinery is design around a CHP gas turbine generator in the range of 2.0 MW and a proprietary WHRU as the primary technologies that reduces Carbon Emission through Circular Energy using a Dual-Fuel system which burns natural gas and / or liquid hydrocarbon. SCR is used to scrub the remaining exhaust flu gas for <GHG.

Specifically designed for Regional Refining of crude oil and Regional Power Generation in third world countries the Turbo Refinery is identified as: CHP+ADU=<GHG.

It's designed and assembles in a *LEGO* type of method, in a standardized small footprint configuration, using ISO standard frames, stackable, module system designed for HD, harsh onshore / offshore applications, and to speed up deployment and improves consistency of on-spec products, while providing multiple product streams from a ~2,500 BPD crude feed stock.

CHP+ADU=<GHG DEFINITION:

CHP stands for: **C**ombined **H**eat and **P**ower

ADU stands for: **A**tmospheric **D**istillation **U**nit

< stands for **less** than

GHG stands for **G**reen **H**ouse **G**ases

Electrical power consumption of the ISBL portion of the unit is approximately 1/3 of the generator capacity leaving the bulk of the generated power for regional use or sale.

136c Smart and Multifunctional Tubacoat Technology for CAPEX and OPEX Reduction, Decarbonization and Biofuels Production in Refining and Petrochemical Applications,

Sanjay Lodha

TUBACOAT is a unique disruptive proven technology for refining, chemical and petrochemical critical applications processing difficult feedstock including biomass, a game changer for the industry for OPEX and CAPEX reduction, decarbonization and biomass processing. This Innovative and customized silica-based inert coating technology from Tubacex Group provides anti-fouling, anti-corrosion and anti-coking properties in harsh environments and extreme temperatures (800°C/1472°F), making TUBACOAT useful in almost all refineries and petrochemical applications, focusing on delayed cokers, visbreakers, cracking, crude vacuum distillation unit furnaces, resid processing, biomass processing, pyrolysis and similar units.

In Most refining conversion units such as delayed coker and others, Unscheduled shutdowns, high maintenance, reduced heat transfer, higher fuel consumption and throughput reduction due to furnace and heat exchanger tube coking, fouling and corrosion is a common problem. TUBACOAT technology minimizes formation, deposition and plugging due to coke/carbon in addition to reduction in fouling, corrosion and fuel usage. Use of TUBACOAT technology in these application leads to longer runs, incremental life cycle for equipment, reliable operations, increased heat transfer, reduced fuel consumptions, reduced CO2 emissions and hence significant savings in OPEX and CAPEX.

Results and case studies from various commercial field trials and runs with major US, Europe, Asia refineries and new applications using Tubacoat technology covered in this presentation. Results include advantages of resistance of ceramic coated tubes to various forms of coking, corrosion, fouling, scaling, plugging, abrasion and the expected cost savings.

Low Temperature Cracking of Low-Density Polyethylene over ZSM-5 Catalysts

Daniel Shantz
Tulane University
6823 St. Charles Avenue, New Orleans, LA 70118
dshantz@tulane.edu

Prepared for Presentation at
American Institute of Chemical Engineers
2024 Spring Meeting and 20th Global Congress on Process Safety
New Orleans, LA
March 24-28, 2024

Low Temperature Cracking of Low-Density Polyethylene over ZSM-5 Catalysts

Daniel Shantz
Tulane University
6823 St. Charles Avenue, New Orleans, LA 70118
dshantz@tulane.edu

Keywords: zeolite, polyethylene, cracking, waste conversion

1 Introduction

The conversion of waste plastics in value added molecules represents a key step towards a more sustainable future. In particular, polyethylene (PE) makes up close to 30% of global plastic production at over 100 Mt in 2015, of which low-density polyethylene (LDPE) is the most widely produced form.[1-3] Unfortunately, most LDPE is used just once and landfilled. Recirculating this carbon back in the chemical/plastics/fuel pool is key for mitigation of landfilling and for mitigating the need to continue increasing oil and natural gas extraction. After capturing the heating value of waste plastic via combustion, pyrolysis and gasification are two of the most established methods of chemical (tertiary) recycling of PE waste, in which the polymer is altered via chemical or thermo-chemical means to produce petrochemical products. More recently numerous reports of hydrocracking waste plastics have appeared.[4-8] While this approach leads to fuel cuts of molecules, the large hydrogen load required to upgrade the LDPE is a potential drawback.

2 Body

In this presentation I will describe my lab's efforts in understanding the feasibility of catalytically cracking LDPE at low temperatures, i.e., below 300 °C, where thermal cracking does not take place.[9] The goal of this research was to determine the effects of solid acid catalyst properties on the cracking of LDPE to establish a methodology for improving its conversion to useful products. The main finding in this work is that it is feasible to crack LDPE at 250 °C to produce light hydrocarbon products at modest rates. The product slate includes methane but also C2 – C4 olefins and paraffins. A key finding was that under these conditions, the zeolite particle morphology generally, and its external surface area more particularly, played a key role. Figure 1 illustrates this for a series of ZSM-5 samples with Si/Al values between 25-50 with different morphologies. As can be seen there is correlation between the external surface and conversion (up to a certain point). SEM images (not shown) are consistent the external surface area values correlation. By contrast, there was no clear correlation between acid site density (Si/Al) and LDPE conversion. Ongoing work is analyzing the various materials via XPS to better understand if aluminum zoning complicates such an analysis. Controls with aluminum free MFI as well as mesoporous silicas such as MCM-41 and SBA-15 showed no LDPE conversion, consistent with the acid sites of the zeolite being responsible for the observed chemistry.

Figure 1. LDPE cracking of several samples of H-ZSM-5 for 8 h at 250 °C with varying external surface area.

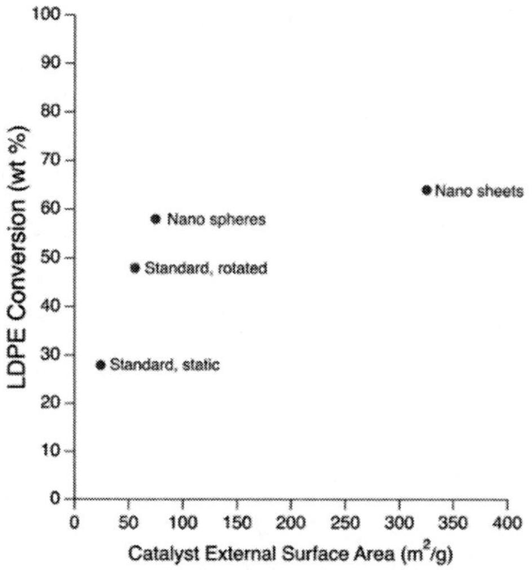

These results will be discussed in the context of how chain branching of the LDPE leads to the inability of the LDPE to diffuse appreciably into the micropore system, leading to what could be referred to as non-classical cracking, wherein the chain ends are selectively cleaved leading to the relatively higher yield of light hydrocarbons observed.

More recent work from our lab will also be described. In this part of the talk, I will discuss our work exploring how the presence of water can influence the reaction rate, production slate, and catalyst stability. This part of the talk will include spectroscopy and other analysis of spent catalysts. A key motivator for this is that preliminary work in our lab indicated a dramatic increase in the LDPE conversion, e.g., from 56 to 88%, in the presence of small amounts of water vapor. I will discuss our preliminary work trying to develop the "why" for the observed behavior.

3 References

1. R. Geyer, J.R. Jambeck, K.L. Law, Sci. Adv. 3 (2017) e1700782. https://doi.org/10.1126/sciadv.1700782.
2. International Energy Agency, Future of petrochemicals report 2018. https://iea.blob.core.windows.net/assets/bee4ef3a-8876-4566-98cf-7a130c013805/The_Future_of_Petrochemicals.pdf (accessed 11 April 2022).
3. United States Environmental Protection Agency, Plastics: material-specific data. https://www.epa.gov/facts-and-figures-about-materials-waste-and-recycling/plastics-material-specific-data (accessed 11 April 2022).
4. D.P. Serrano, J.M. Escola, L. Briones, M. Arroyo, Fuel 206 (2017) 190–198. https://doi.org/10.1016/j.fuel.2017.06.003.
5. A. Marcilla, M.I. Beltrán, R. Navarro, Appl. Catal. B: Environ. 86 (2009) 78–86. https://doi.org/10.1016/j.apcatb.2008.07.026.
6. A.L. Figueiredo, A.S. Araujo, M. Linares, Á. Peral, R.A. García, D.P. Serrano, V.J. Fernandes, J. Anal. Appl. Pyrolysis 117 (2016) 132–140. https://doi.org/10.1016/j.jaap.2015.12.005.

7. A. Galadima, A. Masudi, O. Muraza, J. Environ. Chem. Eng. 9 (2021) 106655. https://doi.org/10.1016/j.jece.2021.106655.
8. D.P. Serrano, J. Aguado, J.M. Escola, J.M. Rodriguez, A. Peral, J. Catal. 276 (2010) 152−160. https://doi.org/10.1016/j.jcat.2010.09.008.
9. J. P. Hittinger, D. F. Shantz, Microporous Mesoporous Mater 343 (2022) 112170. https://doi.org/10.1016/j.micromeso.2022.112170.

SPRING24
+20TH GCPS
A Joint AIChE and CCPS Meeting

Tailoring HZSM-5 for Sustainable Waste Polypropylene Recycling: Enhancing Catalyst Stability and Selectivity in Light Production

Oğuzhan Akin
Laboratory for Chemical Technology (LCT), Department of Materials, Textiles and Chemical Engineering, Faculty of Engineering & Architecture, Ghent University
Technologiepark 125, 9052 Zwijnaarde, Belgium
oguzhan.akin@ugent.be

Qing He
Laboratory for Chemical Technology (LCT), Department of Materials, Textiles and Chemical Engineering, Faculty of Engineering & Architecture, Ghent University
Technologiepark 125, 9052 Zwijnaarde, Belgium
qing.he@ugent.be

Yihan Wang
Laboratory for Chemical Technology (LCT), Department of Materials, Textiles and Chemical Engineering, Faculty of Engineering & Architecture, Ghent University
Technologiepark 125, 9052 Zwijnaarde, Belgium
yihan.wang@ugent.be

Parviz Yazdani
Laboratory for Chemical Technology (LCT), Department of Materials, Textiles and Chemical Engineering, Faculty of Engineering & Architecture, Ghent University
Technologiepark 125, 9052 Zwijnaarde, Belgium
parviz.yazdani@ugent.be

Robin John Varghese
Laboratory for Chemical Technology (LCT), Department of Materials, Textiles and Chemical Engineering, Faculty of Engineering & Architecture, Ghent University
Technologiepark 125, 9052 Zwijnaarde, Belgium
robin.johnvarghese@ugent.be

Kevin M. Van Geem
Laboratory for Chemical Technology (LCT), Department of Materials, Textiles and Chemical Engineering, Faculty of Engineering & Architecture, Ghent University
Technologiepark 125, 9052 Zwijnaarde, Belgium
kevin.vangeem@ugent.be

Prepared for Presentation at
American Institute of Chemical Engineers
2024 Spring Meeting and 20th Global Congress on Process Safety
New Orleans, LA
March 24-28, 2024

—

Tailoring HZSM-5 for Sustainable Waste Polypropylene Recycling: Enhancing Catalyst Stability and Selectivity in Light Production

Oğuzhan Akin
Laboratory for Chemical Technology (LCT), Department of Materials, Textiles and Chemical Engineering, Faculty of Engineering & Architecture, Ghent University
Technologiepark 125, 9052 Zwijnaarde, Belgium
oguzhan.akin@ugent

Qing He, Yihan Wang, Parviz Yazdani, Robin John Varghese, Kevin M. Van Geem
Laboratory for Chemical Technology (LCT), Department of Materials, Textiles and Chemical Engineering, Faculty of Engineering & Architecture, Ghent University

Keywords: Pyrolysis, Polyolefin, Plastic, Catalyst

Abstract

The varying composition of waste polyolefin streams stresses the need for a robust catalyst capable of selectively converting pyrolysis vapors into valuable light olefins while maintaining its stability. HZSM-5, known for efficiently cracking polyethylene pyrolysis vapors, faces challenges in stability during propylene (PP) pyrolysis vapor cracking, due to potential mass transfer limitations and coke formation. This study modified HZSM-5 through phosphorus impregnation, mesopore introduction, and steam treatment to assess their impact on catalyst stability and light olefin selectivity for PP pyrolysis vapor cracking. Sequential experiments revealed the superior stability of modified catalysts, with regeneration experiments showing reversible deactivation from coke formation during pure PP pyrolysis. However, contaminating PP with Polyvinyl chloride (PVC) led to irreversible deactivation for the parent HZSM-5, while P-modified catalysts were successfully regenerated. Phosphorus modification effectively preserved the catalyst framework and acidity under PVC-contaminated conditions, highlighting its effectiveness in developing a resilient catalyst for real-world plastic waste treatment applications.

—

1 Introduction

The rise in plastic production and inadequate disposal practices have resulted in an escalating environmental crisis, prompting concerted efforts to enhance recycling efficiency.[1] Polyolefins, such as polypropylene (PP), polyethylene (PE), and polystyrene (PS), a substantial proportion of the plastic waste encountered in municipal solid waste streams. Effectively recycling polyolefins is imperative for advancing sustainability objectives within the chemical industry and fostering the principles of a circular economy.[2] Conventional disposal methods, notably landfilling and incineration, are fraught with environmental drawbacks, including pollution, greenhouse gas emissions, and the generation of nano-plastics.[3] Consequently, there is a growing interest in chemical recycling strategies, such as pyrolysis, owing to their potential to thermally decompose plastic waste into valuable products, offering scalability and operational versatility in industrial settings.

Pyrolysis enables the thermolytic breakdown of polyolefins into lower-value oils[4], which can be suitable for integration into steam-cracking processes provided they meet specific quality standards.[5] Nevertheless, conventional thermal pyrolysis necessitates elevated temperatures and substantial energy inputs, thereby constraining its widespread adoption. This process entails multiple stages, encompassing the conversion of complex pyrolysis oils (ranging from C_5 to heavy waxes)[6], subsequent purification steps, and steam cracking with an average monomer recovery efficiency of 70%.[7] In contrast, catalytic pyrolysis has emerged as a promising alternative route. This recycling approach not only reduces energy demands and process intricacy but also enhances the selectivity towards desirable products, such as light olefins (C_2-C_4 olefins), building blocks of various plastic materials and chemical compounds.[8] Enhanced selectivity for light olefins can be achieved either through ex-situ upgrading of thermal pyrolysis volatiles in a secondary reactor closely coupled to the primary pyrolysis unit or via in-situ processing, where the catalyst is directly mixed with plastic waste and introduced into a thermo-catalytic pyrolysis reactor. Ex-situ catalytic pyrolysis presents a notable advantage over its in-situ counterpart when considering the presence of non-volatile contaminants in plastic waste. By initially subjecting the waste to thermal pyrolysis, a portion of non-volatile contaminants is prevented from directly interacting with the catalyst, thereby mitigating the risk of catalyst deactivation due to significant char formation.[9] However, hazardous substances within the plastic waste, such as those generated from PVC, remain a significant concern for ex-situ catalytic pyrolysis. The pyrolysis of PVC yields highly corrosive and toxic HCl gas, which can irreversibly deactivate the catalyst through dealumination.[9] Thereby, the recycling efficiency of PP waste is hampered by its heterogeneous compositions and varying contamination levels.

Product distribution and selectivity hinge on various factors, including pyrolysis temperature, catalyst type, feedstock characteristics, the residence time of pyrolysis vapors in the reactor, and the contact time with catalysts.[10] For instance, high pyrolysis temperatures (above 600 °C) result in increased yields of gaseous products and coke formation in the presence of a catalyst. Simultaneously, a short catalyst-vapor contact time helps restrict secondary reactions leading to aromatic formation but limits the cracking of

—

heavy pyrolysates.[8] Microporous catalysts, such as protonated Zeolite Socony Mobil–5 (HZSM-5), are renowned for their efficacy in catalyzing the cracking of polyethylene (PE) pyrolysis vapors, which primarily consist of linear aliphatic hydrocarbons. In contrast, polypropylene pyrolysis vapors predominantly contain branched olefins.[11] The catalytic cracking of these branched olefins is substantially influenced by the mass transfer constraints within the micropores of the catalyst, necessitating additional modifications to enhance catalyst performance.

Our previous study has shown the excellent selectivity to light olefins and high stability against coke formation of severely-steamed phosphorous-modified HZSM-5 during the pyrolysis vapor cracking of mixed polyolefin waste (MPO), which predominantly contained PE (75wt.% PE), 16wt. % PP, 2 wt.% PA, 1 wt.% PET, 1 wt.% PS and 5 wt.% non-polymers).[12] Thus, the product yields of MPO pyrolysis were comparable to virgin HDPE. Given the widespread utilization of PP across various sectors such as packaging, construction, and medical equipment, it is imperative to explore the performance of PP pyrolysis vapor cracking over phosphorus-modified HZSM-5 for the development of robust catalysts for polyolefin recycling technologies. Notably, the existing literature lacks comprehensive investigations into the production of light olefins from polypropylene via direct catalytic conversion of pyrolysis vapors, with studies predominantly focusing on the generation of low-value pyrolysis oil and aromatics rather than light olefins.

In this work, we propose modifying phosphorus-impregnated HZSM-5 with mesopores to facilitate the diffusion of bulky branched olefins into the catalyst pores, overcoming potential diffusion limitations to maximize light olefin selectivity and improve the stability. Hence, catalytic upgrading of virgin PP and PVC (5 wt.%) containing PP pyrolysis vapors were investigated using the micro-pyrolyzer facility combined with comprehensive two-dimensional gas chromatography (GCxGC-FID/TOF-MS). This analytical approach enables accurate quantification and identification of complex pyrograms, shedding light on the intricate process of pyrolysis and catalyst performance.

2 Materials and Methodology

2.1 Materials

Polypropylene (PP) and high-density polyethylene (HDPE) samples were received from Borealis (HE125MO) and Dow (HDPE 25055E), respectively. Samples were obtained as pellets and downsized using a cutting mill (FRITSCH). The produced powder was sieved to obtain a fraction with particle size < 300 μm to avoid mass and heat transfer limitations during micro-pyrolysis. The HZSM-5 (parent catalyst) with nominal SiO_2/Al_2O_3 ratio 55 was obtained from Zeolyst International (sample code CBV 5524 G).

2.2 Catalyst Synthesis

HZSM-5 was transformed from the as-received ammonium form to the proton form through calcination in dry airflow at 550 °C for 5 h. Mesopores introduction to parent

HZSM-5, involved desilication with an NaOH solution followed by mild acid wash, as described in detail in Eschenbacher et al.[12]

The phosphorus loading was maintained at 2 wt.%, and the catalysts were subjected to severe steaming at 800 °C for 4 hours, as it exhibited the most favorable stability and activity for the cracking of mixed polyolefin pyrolysis vapor, specific details about phosphorous impregnation and steaming can be found elsewhere.[12] The steamed phosphorus-modified HZSM-5 is abbreviated as P-HZSM5, while steamed phosphorus-modified mesopore-introduced HZSM-5 as P-mHZSM5 hereon. Before catalytic pyrolysis tests, all catalysts were ground and sieved to obtain a particle size range of 150-300 μm.

2.3 Catalyst Characterization

N_2 physisorption was carried out at liquid nitrogen temperature on a Micromeritics Tristar II adsorption analyzer according to ASTM D4365. Total acidity was determined by NH_3-TPD using an Autochem II instrument (Micrometrics). Details about NH_3-TPD analysis can be found elsewhere.[8]

2.4 Micropyrolyzer

The experiments were conducted by using a single-shot tandem micro-pyrolysis facility (Rx-3050tr, Frontier Labs, Japan) coupled with two-dimensional gas chromatography (GC×GC) and a separate customized multicolumn GC (Trace 1300) for the analysis of light gases (Figure 1). The unit consists of a pyrolysis reactor and catalytic reactor containing a fixed bed. He was used as carrier gas at variable flow rates of 50 mL/min purge volatiles in pyrolysis reactor into the second catalytic reactor. The catalyst to plastic (C/P) ratio was kept as 80 (wt./wt.) for P-HZSM5 and P-mHZSM5 and 20 (wt./wt.) for the parent HZSM-5 due to its significantly higher acidity. Parent HZSM-5 was mixed with inert α-alumina to have a similar bed height with modified catalysts (10 mm).

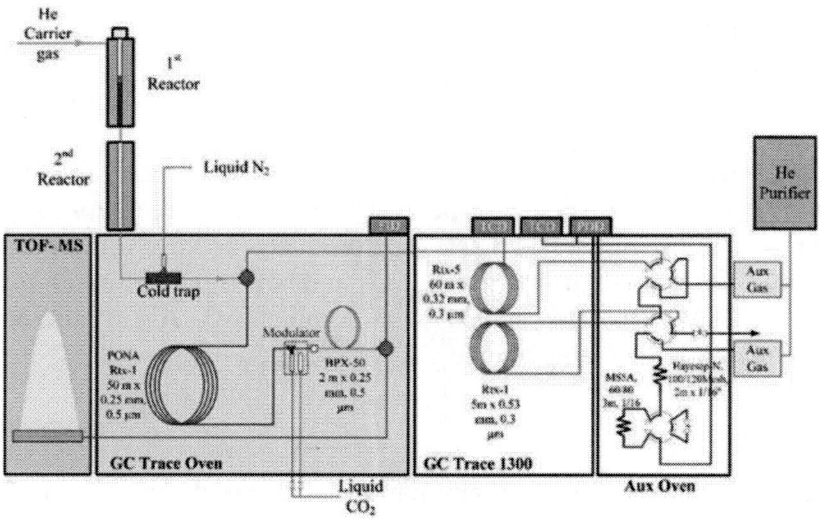

Figure 1. Micropyrolysis unit [13].

—

3 Results and Discussion

3.1 Catalyst Properties

The physicochemical properties of catalysts were investigated in our previous publication[12] and are summarized in Table 1. Both modified catalysts demonstrated lower BET surface area compared to parent HZSM-5. Moreover, steaming significantly decreased the total and Brønsted acidity, which is highly active for secondary reactions leading to aromatics during olefin cracking.[14]

Table 1. The physicochemical properties of the studied catalysts.[12]

Property	HZSM-5	P-HZSM5	P-mHZSM5
Acidity[mmol NH_3/g]	0.65	0.1	0.09
V_{meso} [cm^3/g]	0.09	0.06	0.11
V_{total} [cm^3/g]	0.31	0.3	0.43
BET [m^2/g]	423	401	349
Brønsted acid sites [mmol/g]	0.26	0.05	0.02
Lewis acid sites [mmol/g]	0.04	0.02	0.01
Brønsted/Lewis acidity	6.5	2.5	2

3.2 Thermal Pyrolysis

The thermal pyrolysis of PP was investigated initially to determine the product yields and establish a reference for assessing the influence of catalyst on product distribution. Thermal pyrolysis at 600 °C resulted in 25 wt.% total C_2-C_4 light olefin yield with a dominant propylene contribution (15 wt.%). The fingerprint products for PP pyrolysis were oligomers of propylene: 1-Pentene, 2-methyl- (dimer), 2,4-Dimethyl-1-heptene (trimer), 2,4,6-trimethylnon-1-ene (tetramer), and tetramethylene-1-ene (pentamer) of which the major product was trimer (10 wt.%).[15] Compared to HDPE, PP pyrolysis leads to higher light olefins selectivity (25 wt.% vs. 9 wt.%) and higher branched olefin yields (excluding di-olefins), 43.3 wt.%, as demonstrated in Figure 2. This can be attributed to the structural difference between the two polyolefins. PP chains have a larger cross-sectional area than PE owing to the presence of methyl group substituents on alternating carbon atoms in the chain. These methyl groups increase the number of tertiary carbons having considerably lower resistance to degradation in the molecule. As a result, thermal cleavage of C-C bonds occurs more readily, leading to the formation of methyl, primary, and secondary alkyl radical.[16]

—

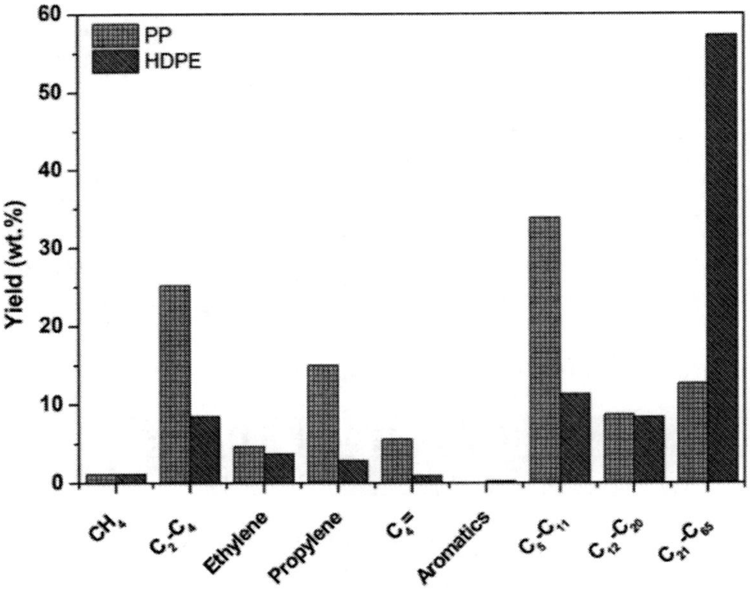

Figure 2. Products obtained from thermal pyrolysis of PP and HDPE at 600 °C.

3.3 Catalytic Upgrading of Polypropylene Pyrolysis Vapors

Catalytic performance of parent HZSM-5, P-HZSM5, and P-mHZSM5 was tested at 600 °C for selective light olefin production for upgrading PP pyrolysis vapors formed at 550 °C. All the tested catalysts have demonstrated excellent activity for cracking of propylene trimer and yielded enhanced light olefin selectivity. Figure 3 demonstrates polypropylene pyrolysis vapor cracking product yields over HZSM-5, P-HZSM5, and P-mHZSM5 at 600 °C for a single run. The highest light olefin selectivity was obtained with HZSM-5 (79 wt.%). The number of acid sites present in the catalyst bed loaded with parent HZSM-5 is ~1.5 times higher than that of P-HZSM5 and P-mHZSM5, even after diluting with α-alumina. Our previous study also observed a similar trend for the catalytic cracking of mixed polyolefin pyrolysis. P-HZSM5 and P-mHZSM5 yielded similar selectivity for light olefins, 74 wt.% and 75 wt.%, respectively. Given that both catalysts have similar acidities, introducing mesopores seems to have a slight influence on light olefin selectivity for a single run.

—

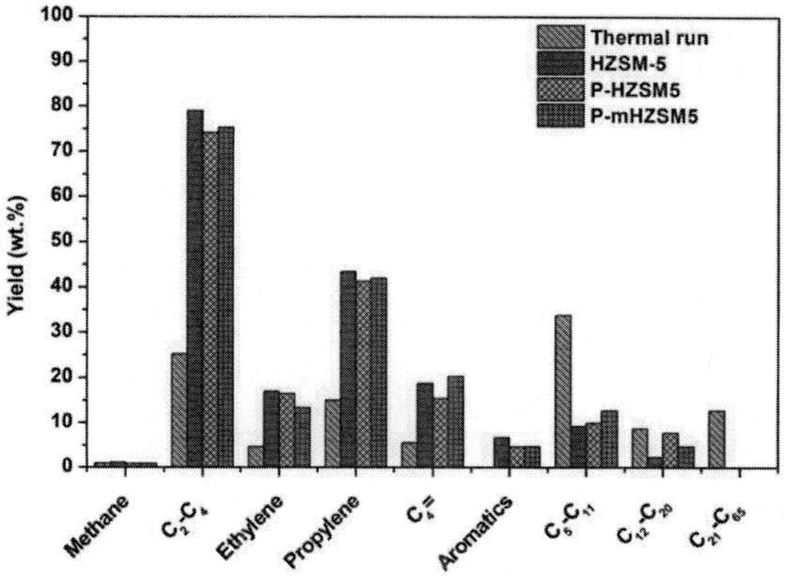

Figure 3. Product yields of polypropylene pyrolysis vapor cracking over HZSM-5, P-HZSM5, and P-mHZSM5 at 600 °C for a single run.

3.4 Catalytic Upgrading of Polypropylene Pyrolysis Vapors

Despite exhibiting superior performance in an initial single trial, the modified catalysts surpassed the parent catalyst, due to modification that resulted in the removal of strong acid sites and a reduction in total acidity, as well as the Brønsted/Lewis acid site ratio. Consequently, this led to a significant reduction in their susceptibility to coke deposition, enabling sustained activity over prolonged operational periods, as depicted in Figure 4. Throughout a series of 150 experiments, both modified catalysts consistently outperformed the parent catalyst. Notably, the parent catalyst, characterized by its higher activity and propensity for aromatic formation, displayed the most pronounced decline in C_2-C_4 olefin selectivity during the initial 50 runs, decreasing from 79 wt.% to 60 wt.%. Subsequently, its deactivation continued over the subsequent 100 runs, with C_2-C_4 olefin selectivity reduced to 44 wt.%. Similarly, the P-HZSM5 catalyst experienced a notable decrease in C_2-C_4 olefin selectivity during the first 50 runs, dropping from 74 wt.% to 63 wt.%. However, it demonstrated sustained activity for the subsequent 50 runs and only demonstrated a decrease at the end of 150th run, yielding 57 wt.% C_2-C_4 olefins. Conversely, the incorporation of mesopores into the P-HZSM5 catalyst enhanced its stability and reduced the rate of coke formation, with the P-mHZSM5 catalyst exhibiting the most favorable stability profile over 150 runs. For instance, light olefin yields decreased to only 63 wt.% after 100 runs and remained consistent for the subsequent 50 runs, representing the highest selectivity among the tested catalysts. This observed phenomenon can be attributed to the combined effect of mesopore introduction and a diminished number of Brønsted acid sites. Notably, coke formation on the catalyst need not be viewed solely as detrimental, as the exothermic combustion of coke during regeneration is integral to heat management of the overall process and promoting autothermal operation. Consequently, catalysts were successfully regenerated via coke combustion with air at 700 °C,

—

demonstrating that deactivation was not irreversible, as all regenerated catalysts exhibited similar C_2-C_4 olefin yields to their fresh counterparts.

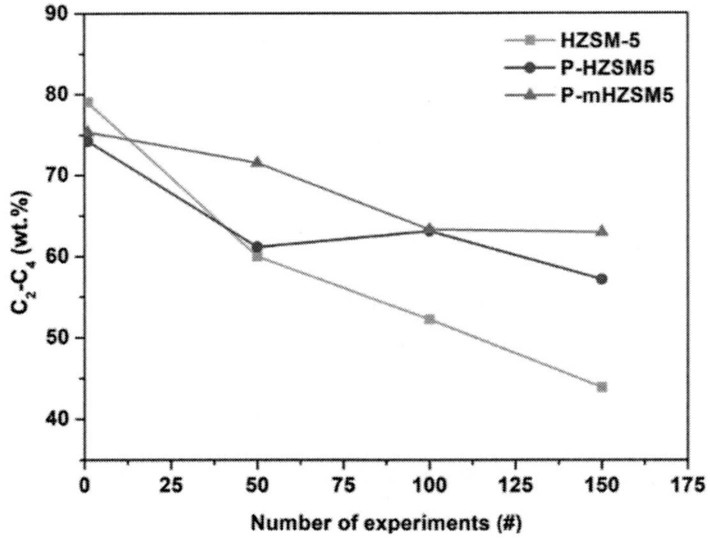

Figure 4. C_2-C_4 yields for increasing number of consecutive experiments using HZSM-5, P-HZSM5 and P-mHZSM5 catalysts at 600 °C.

In order to evaluate the catalyst performance under conditions more representative of real-world applications, the polypropylene feedstock was intentionally contaminated with PVC (5 wt.%). The deactivation trends and comparisons of freshly regenerated catalysts are presented in Figure 5. Notably, in the case of the parent catalyst, the degree of deactivation was less pronounced compared to pure virgin PP pyrolysis, resulting in a C_2-C_4 olefin yield of 65 wt.% after 50 runs. This observed phenomenon can be attributed to the release of HCl gas during PVC pyrolysis, which has higher tendency to deactivate the weak acid sites responsible for bimolecular reactions, thus keeping strong acid sites remained active for olefin cracking. However, the deactivation observed after 50 runs for the modified catalysts was considerably more severe compared to the virgin PP pyrolysis case, with C_2-C_4 yields dropping to ~55 wt.%. The absence of complete coke formation within the initial 50 runs suggests potential poisoning of active sites by HCl. Due to its small molecular size, HCl can readily penetrate the micropores of catalysts and form chlorine-containing intermediates with olefins, thereby impacting catalytic activity and selectivity.

Catalyst regeneration revealed intriguing outcomes. While both the P-modified and mesoporous P-modified catalysts were effectively regenerated, the parent catalyst underwent irreversible deactivation. Notably, the decoking process exacerbated the deactivation of the parent catalyst, resulting in a reduction of ~4 wt.% in C_2-C_4 olefin yields compared to its coke-laden state. This phenomenon can be attributed to the combustion of chlorine-containing coke on the catalyst surface, leading to further dealumination and framework rupture, thereby reducing the number of active sites. In contrast, the interaction between phosphorus and partially displaced framework alumina can enhance the stability of the zeolite framework by forming localized framework silico-aluminosphosphate

—

(SAPO) interfaces.[17] These interfaces secure Si-O-Al bonds within the zeolite lattice, thereby preserving the Si/Al ratio and pore structure. This elucidates the heightened stability of P-modified catalysts against dealumination, a phenomenon potentially induced by the release of HCl-containing vapors during PVC-containing PP pyrolysis, ultimately leading to successful catalyst regeneration.

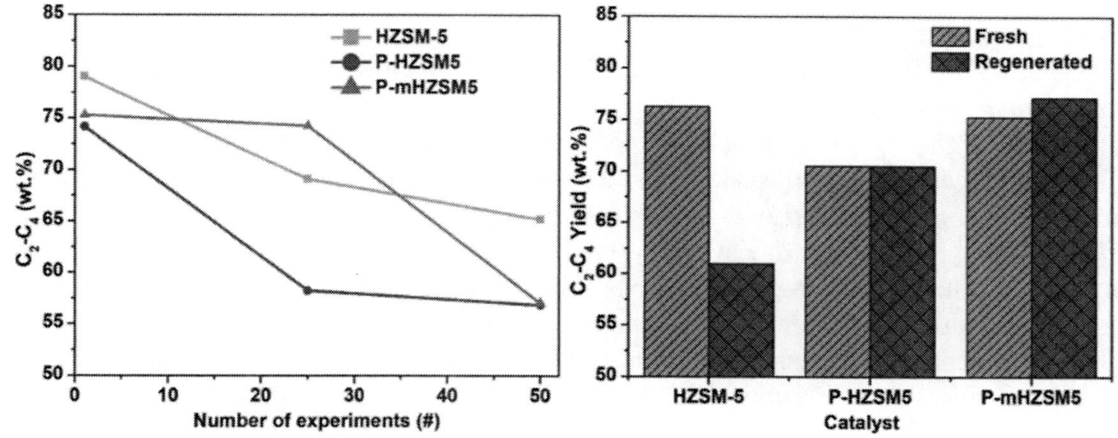

Figure 5. C_2-C_4 yields for increasing number of consecutive experiments using HZSM-5, P-HZSM5 and P-mHZSM5 during PVC containing PP pyrolysis (left), comparison of C2-C4 yields for fresh and regenerated catalysts (right).

4 Conclusion

Polypropylene was effectively converted into light olefins with high selectivity using parent, P-modified, and mesoporous P-modified HZSM-5 catalysts. P-impregnation with mesopore introduction enhanced light olefin selectivity and stability against coke formation for polypropylene pyrolysis. Deactivation due to coke formation during pure PP pyrolysis vapor cracking over 150 runs was reversible, as all the catalysts were successfully regenerated. Pyrolysis of PVC-contaminated PP led to even faster deactivation for all catalysts, with light olefin selectivity dropping to below ~65 wt.% even after 50 runs. However, the P-modified catalysts were effectively regenerated, whereas the parent HZSM-5 exhibited permanent activity loss. This is possibly originating from P-modification preventing dislodged alumina from leaving the zeolite framework by interacting with Si and leached Al to form SAPO species, thereby preserving the framework and acidity. Nevertheless, further analysis on spent catalysts is required to better understand the mechanism of phosphorus influence on the catalyst stability. Considering the significant impact of 5 wt.% PVC contamination in the feed, the development of robust catalysts is essential for scaling up the proposed plastic waste recycling route, where P-impregnation has demonstrated promising results.

—

5 References

[1] Eschenbacher A, Varghese RJ, Abbas-Abadi MS, Van Geem KM. Maximizing light olefins and aromatics as high value base chemicals via single step catalytic conversion of plastic waste. *Chemical Engineering Journal.* 2022;428:132087.
[2] Van Geem KM. Plastic waste recycling is gaining momentum. *Science.* 2023;381(6658):607-608.
[3] Ragaert K, Delva L, Van Geem K. Mechanical and chemical recycling of solid plastic waste. *Waste Manage.* 2017;69:24-58.
[4] Dogu O, Pelucchi M, Van de Vijver R, et al. The chemistry of chemical recycling of solid plastic waste via pyrolysis and gasification: State-of-the-art, challenges, and future directions. *Progress in Energy and Combustion Science.* 2021;84.
[5] Kusenberg M, Roosen M, Zayoud A, et al. Assessing the feasibility of chemical recycling via steam cracking of untreated plastic waste pyrolysis oils: Feedstock impurities, product yields and coke formation. *Waste Manage.* 2022;141:104-114.
[6] Eschenbacher A, Goodarzi F, Varghese RJ, et al. Boron-Modified Mesoporous ZSM-5 for the Conversion of Pyrolysis Vapors from LDPE and Mixed Polyolefins: Maximizing the C-2-C-4 Olefin Yield with Minimal Carbon Footprint. *Acs Sustain Chem Eng.* 2021;9(43):14618-14630.
[7] Lange JP. Towards circular carbo-chemicals - the metamorphosis of petrochemicals. *Energ Environ Sci.* 2021;14(8):4358-4376.
[8] Akin O, Varghese RJ, Eschenbacher A, et al. Chemical recycling of plastic waste to monomers: Effect of catalyst contact time, acidity and pore size on olefin recovery in ex-situ catalytic pyrolysis of polyolefin waste. *Journal of Analytical and Applied Pyrolysis.* 2023;172.
[9] Dutra C, Freire MTD, Nerin C, et al. Migration of Residual Nonvolatile and Inorganic Compounds from Recycled Post-Consumer PET and HDPE. *J Brazil Chem Soc.* 2014;25(4):686-696.
[10] Abbas-Abadi MS, Ureel Y, Eschenbacher A, et al. Challenges and opportunities of light olefin production via thermal and catalytic pyrolysis of end-of-life polyolefins: Towards full recyclability. *Progress in Energy and Combustion Science.* 2023;96.
[11] Kusenberg M, Zayoud A, Roosen M, et al. A comprehensive experimental investigation of plastic waste pyrolysis oil quality and its dependence on the plastic waste composition. *Fuel Process Technol.* 2022;227.
[12] Eschenbacher A, Varghese RJ, Delikonstantis E, et al. Highly selective conversion of mixed polyolefins to valuable base chemicals using phosphorus-modified and steam-treated mesoporous HZSM-5 zeolite with minimal carbon footprint. *Appl Catal B-Environ.* 2022;309.
[13] SriBala G, Vargas DC, Kostetskyy P, et al. New Perspectives into Cellulose Fast Pyrolysis Kinetics Using a PyGC x GC-FID/MS System. *Acs Eng Au.* 2022;2(4):320-332.
[14] Olazar M, Lopez G, Amutio M, Elordi G, Aguado R, Bilbao J. Influence of FCC catalyst steaming on HDPE pyrolysis product distribution. *Journal of Analytical and Applied Pyrolysis.* 2009;85(1):359-365.

—

[15] Kusenberg M, Eschenbacher A, Djokic MR, et al. Opportunities and challenges for the application of post-consumer plastic waste pyrolysis oils as steam cracker feedstocks: To decontaminate or not to decontaminate? *Waste Manage.* 2022;138:83-115.
[16] Kassargy C, Awad S, Burnens G, Kahine K, Tazerout M. Experimental study of catalytic pyrolysis of polyethylene and polypropylene over USY zeolite and separation to gasoline and diesel-like fuels. *Journal of Analytical and Applied Pyrolysis.* 2017;127:31-37.
[17] Ibanez M, Gamero M, Ruiz-Martinez J, et al. Simultaneous coking and dealumination of zeolite H-ZSM-5 during the transformation of chloromethane into olefins. *Catalysis Science & Technology.* 2016;6(1):296-306.

153c Tandem Dechlorination and Hydrogenation of PVC to Waxes.

Konstantinos Goulas, Scott Svadlenak

Polyvinyl chloride (PVC) recycling rates are far below than those of polyolefins, such as polyethylene and polypropylene. Moreover, chemical recycling processes are incompatible with PVC, due to the presence of chlorine in the molecule. These concerns spur us to develop an upcycling methodology to profitably upcycle waste PVC to a wax product, and sequester chlorine as sodium chloride salt. Our proposed process proceeds via a tandem dechlorination and hydrogenation process. In this context, the dechlorination is catalyzed by a base and takes place homogeneously. The poly-ene macromolecule that is formed by the dechlorination is then hydrogenated to a polyethylene-like molecule over a heterogeneous catalyst, such as Pt or Pd. The dechlorination is driven by the partitioning of amine bases into the polymer phase, while the hydrogenation competes with C-C bond scission over the heterogeneous catalyst, as well as homogeneous char formation. By tuning the reaction conditions, we are able to suppress side reactions and produce waxes at 78% yield.

184a CFD Modeling of Industrial Burners: Investigation of Hydrocarbon Fuel Vs. Carbon-Free Ammonia Based Fuel

Krishnadash S. Kshetrimayum, Minkyu Han, Younghee Lee

In the present time, amid the global effort to fight climate change, research related to the use of Green Ammonia based fuel as an alternative candidate for carbon-free fuel is increasing. In this presentation, we will present the modeling process of a conventional industrial burner used for a specific industrial application, considering the detailed design of the burner and the involved combustion chemistry and physics. The results of combustion characteristics, in terms of flame temperature, laminar burning velocity, flame pattern and exhaust gas composition, using conventional hydrocarbon fuel will be first discussed. Thereafter, results from combustion simulation of Ammonia based fuel (blended with different % of Hydrogen) considering the combustion chemistry of Ammonia will be discussed including the modeling challenges experienced. Understanding gained is expected to help in the design and operation of new burners for applications in new processes related to energy transitions, such as Ammonia Cracking, as well as those of other existing processes in different industrial sectors to play a part in the present decarbonizing effort.

184b Process Design, Simulation and Techno-Economy Analysis of the Green Ammonia Production from Produced Water

Lucylitzy Bautista, Ana Gonzalez, Vahid Taghikhani, Mohammad Moosazadeh

It has been reported that 50 million barrels of produced water are coming out every day from approximately 900,000 oil wells in the United States. The produced water, PW, is highly toxic containing mineral salts, heavy and rare earth metals and can be a major source of water contamination when re-injected underground. Attempts have been made to treat PW through various technologies developed for water desalination over the past few years. The supercritical water and oxidation (SCWDO) technology, recently developed at Los Alamos National Laboratory, has been one of the novel technologies introduced to produce drinking water from PW. AspenHysys as the most popular and commonly used process simulation tool, has been utilized to design various hydrogen production processes. Recently, the process design has been programmed to use flare gas to produce blue hydrogen from the steam methane reforming, SMR reaction, and PW which would be considered as a great advancement towards decarbonization and net zero policy by 2050. In this research proposal, AspenHysys V.14, will be used to design a process to produce green ammonia from PW. Ammonia, as a strategic chemical, is widely used in various industries such as agricultural and medical. The current market value of ammonia has been reported to be 75.7 billion U.S. dollars in 2022 and is anticipated to reach 103.8 billion U.S. dollars in 2028. Ammonia has recently been used as a clean fuel and source of energy in power and aviation industries. It should be stated that hydrogen storage and transmission need careful safety considerations and hazard analysis which make the hydrogen utilization more challenging. Using hydrogen carrier has shown to be an alternating technique for hydrogen storage and transmission. Therefore, chemical conversion of hydrogen to ammonia can be considered as one of the safest and most commercially viable methods. To do so, a produced water treatment, PWT, unit will be designed using real data collected from typical oil wells in Texas or other states in the US. Based on the real compositional data from the collected samples, the water treatment and desalination unit, WTU, will be designed. The effluent hydrogen from WTU will be fed to a Water Splitting Unit, WSU, which will be designed based on a renewable source of energy for the electrolysis process. Water Splitting technology has been proven to produce green hydrogen. Green hydrogen, salts, and critical minerals will leave WSU, and the green hydrogen will be used in an Ammonia Production Plant, APP, to produce green ammonia. The Haber-Bosch process, also called Haber ammonia process, or synthetic ammonia process, will be simulated at high pressure and relatively low temperature. The lower the temperature and the higher the pressure used, the greater the proportion of ammonia yielded in the mixture. Techno-Economy analysis for the proposed process will be performed and the total costs per ton of ammonia will be determined. The results of this process design research can be used by petrochemical, refinery, and other chemical plants to take actions towards green technologies and sustainability in chemical technologies.

184c Evaluating Alternative Energy Sources

Ian Sutton

Currently around 85% of the world's energy is provided by fossil fuels: coal, oil and natural gas. The consequences to the climate of burning these fuels are increasingly serious and self-evident. Therefore, the process industries are under pressure to quickly transition to alternative, 'clean' sources of energy. (The year 2050 is frequently suggested as being a target for reducing greenhouse gas emissions to 'Net Zero'.)

Proposed replacement energy sources include solar (photovoltaic and thermal), wind, nuclear (fission), hydrogen, geothermal, ocean (wave and tidal), biomass, and hydroelectric. Each of these energy sources has many positive characteristics. However, none of them offer the unique combination of features provided by the fossil fuels, particularly oil. (Hydrogen is not a source of energy, but it is included because its use as an energy carrier.)

This paper reviews the different energy sources, and assesses their strengths and limitations.

The Table provides a means of evaluating alternatives. The top row shows the energy sources. The first column in the Table lists the features of each alternative that can help assess its feasibility as a replacement for the fossil fuels.

	Oil	**Nuclear**	**Solar**	**Wind**	**Hydrogen**	**Geothermal**	**Ocean**	**Biomass**	**Hydroelectric**
Energy Dense	Y	Y	N	N	P	N	N	Y	Y
Renewable	N	N	Y	Y	N	P	Y	Y	Y
Available	Y	Y	P	P	Y	Y	N	Y	Y
Dispatchable	Y	Y	N	N	Y	N	N	Y	P
Portable	Y	Y	N	N	Y	N	Y	Y	N
Safe	Y	P	Y	Y	P	Y	Y	Y	Y
Scalable	Y	N	Y	Y	Y	P	Y	Y	N
Petrochemical Source	Y	N	N	N	P	N	N	Y	N
GHG Emissions	Y	N	N	N	N	N	N	N	N
Environmental	P	N	N	P	N	N	P	N	P

The first column lists the properties of each energy source. An evaluation for each energy type is then provided (Y is 'Yes', *i.e.,* desirable, N is 'No', P is 'Partial').

Unfortunately, there is no single source of energy that offers the same combination of features as the fossil fuels, particularly crude oil. For example,

- Nuclear is not dispatchable (it cannot be ramped up or down quickly).
- Solar and wind are not always available (the sun does not shine for half the day, nor does the wind blow when wanted).
- Geothermal is not scalable — there is a limit to how widespread its use can be.
- Biomass has a huge environmental impact because so much land is needed to grow the raw materials.

This means that the future energy grid will be more complex than today's because there will be many energy sources. (The grid will also have to include energy storage.) The manner in which they can be fitted into an overall grid is discussed using the following sketch.

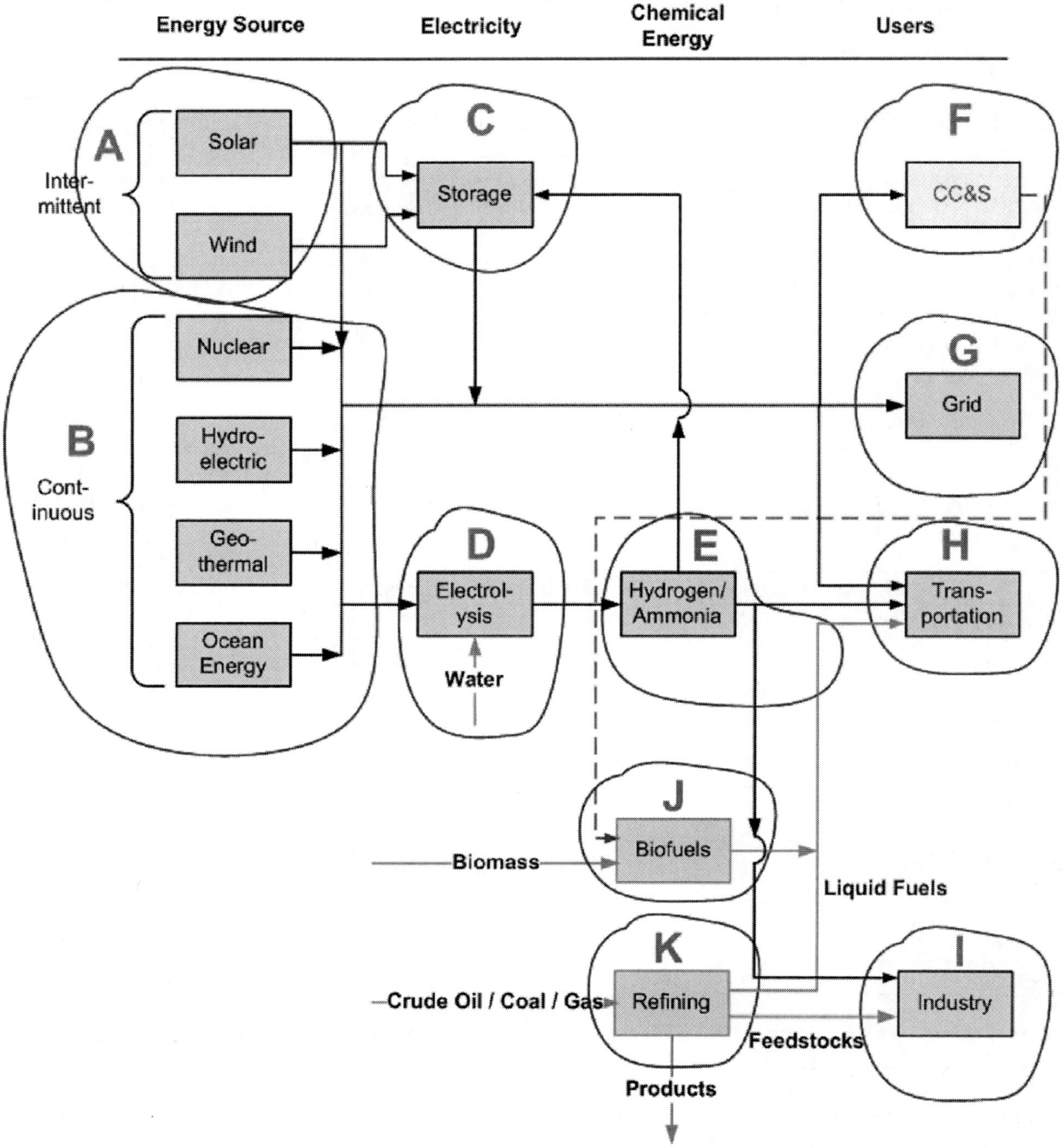
Energy Source
Electricity
Chemical Energy
Users
A
Inter-mittent
Solar
Wind
B
Cont-inuous
Nuclear
Hydro-electric
Geo-thermal
Ocean Energy
C
Storage
D
Electrol-ysis
Water
E
Hydrogen/ Ammonia
F
CC&S
G
Grid
H
Trans-portation
J
Biofuels
Biomass
Liquid Fuels
K
Refining
Crude Oil / Coal / Gas
I
Industry
Feedstocks
Products

96a Bridging Planning, Scheduling, and Real-Time Operations: Expanding the Horizon of APC with Multiperiod Optimization

Filippo Trivella

The paper presents a transformative technology redefining the process industry's Advanced Process Control (APC) boundaries through multiperiod optimization.

This innovative approach consolidates planning, scheduling, and real-time operations into a cohesive optimization framework, facilitating swift adaptations to fluctuating market conditions and increasing operational resilience. Enhancing adaptability, the methodology ensures synchronized orchestration across various process units, promoting just-in-time production and strategic maneuverability in response to process changes, from blending lineups to feed composition switches.

Users can harness models from pre-existing multivariable predictive control (MPC) applications to illustrate linear interrelationships between variables. This approach capitalizes on the substantial investments previously made in advanced process control. Additionally, users have the flexibility to create custom models, employing both linear and non-linear equations to define relationships between variables. For instance, non-linear blending functions can be applied to tank compositions, while reactor models can utilize non-linear yield functions. Furthermore, the model possesses the capability to interface with the planning and scheduling system. This facilitates the integration of financial values, updates on present and anticipated tank line-ups, as well as modifications in operational modes.

With its local and short-term optimization focus, conventional APC often falls short in addressing broader operational challenges. In contrast, this technology boldly broadens the optimization horizon and significantly expands its scope. This allows us to obtain more benefits than conventional APC applications.

The increased scope and horizon allow for the inclusion of planning and scheduling objectives and constraints that can be solved simultaneously.

The paper includes practical case studies in refinery kerosene blending and ethylene cracker units that clearly illustrate this optimization methodology's efficacy and strategic superiority.

SPRING24 +20TH GCPS

A Joint AIChE and CCPS Meeting

Use Material Balance Equations to Solve Difficult Level Control Problems

Charles D. Herzog
Retired Chemical Engineer
Katy, Texas
chasherz@aol.com

Prepared for Presentation at
American Institute of Chemical Engineers
2024 Spring Meeting and 20th Global Congress on Process Safety
New Orleans, LA
March 24-28, 2024

Use Material Balance Equations to Solve Difficult Level Control Problems

Charles D. Herzog
Retired Chemical Engineer
Katy, Texas
chasherz@aol.com

Keywords: equation-based level control, material balance, drums-in-series, inventory control

Abstract

This paper illustrates the direct use of material balance equations for liquid level control, and other inventory control applications, while avoiding the inherent deficiencies of DCS-based proportional-integral (P-I) level control. (Fig. 1)

The advantages of equation-based level control are:

a. Accumulation control via equal outflow moves following feed rate change
b. Return of the level to set point within specified time, with drum in material balance
c. Simultaneous control of drums in series for best response to disturbances
   1. Ethylene plant demethanizer feed system and demethanizer bottoms
   2. Distillation feed drum + reflux drum or bottoms holdup
d. Increased residence time to dampen composition variations without risk of overfill
e. Improved control of systems with dead time between level and valve
f. Pressure control and partial pressure control of vapor systems via material balance

With equation-based control, the user determines the allowable accumulation (ΔV_A) for typical feed rate changes. Allowable accumulation may be a constant, or an exponential function of the magnitude of the disturbance. Using the proper exponent (less than 1) allows larger disturbances to have somewhat greater accumulation. With P-I control, by comparison, accumulation is always directly proportional to the magnitude of the disturbance.

1 Introduction—Why use Equation-Based Level Control?

Most vessels in oil refineries and petrochemical plants use the DCS-based P-I control algorithm for maintaining liquid levels at or near set point. Sometimes, large feed rate disturbances lead to problems such as overfill or emptying, or unsteady outflow to sensitive downstream units. The problem becomes more difficult if the vessel is small relative to the magnitude of the feed rate change.

Even if a disturbance does not lead to overfill or emptying, the return to set point may involve overshoot and oscillation, affecting sensitive downstream units. The level and outflow finally become steady via decreasing wave amplitude. If the P-I loop is poorly tuned, oscillations may continue long after the disturbance has passed, even forcing the operator to take manual control.

Such was the case in an ethylene plant in 1982, where decoking-related feed rate changes routinely caused major upsets in the recovery section of the plant. As the APC engineer at the plant, I developed equation-based level control and drums-in-series control for the demethanizer feed system and demethanizer bottoms, completely eliminating this problem.

Use of equation-based level control is illustrated in this and several other applications, including distillation system feed and product drums, vapor-phase reactor control, and a level control application with several minutes of dead time.

2 Equation- Based Level Control Implementation

The major objectives of level control for vessels in continuous processes (Fig. 1) include:

a. Prevent liquid overfill or emptying
b. Directly limit accumulation following a disturbance
c. Minimize variation of outflow to downstream units
d. Return level to set point without going beyond, and without oscillation

Fig. 2 displays the relationship between inflow, outflow, and accumulation.

Level control is a 3-step process:

1. Equalize outflow with inflow
2. Create small imbalance to return level to set point within specified time
3. Eliminate the imbalance as the level reaches the set point.

Each of these steps is discussed below.

2.1 Equalize Vessel Outflow with Inflow (Accumulation Control)

When liquid feed rate to a vessel changes (inflow disturbance), an imbalance between inflow and outflow is created. This imbalance causes accumulation, which is measured as a change in the liquid level. As shown in Eq. 1a, accumulation (ΔV) is the integral of the imbalance over time.

$$\Delta V = \int (F_{IN} - F_{OUT})\, dt \quad (1a)$$

$$\Delta V = \text{Liquid Accumulation} = \text{Average Imbalance x Time} \quad (1b)$$

$$B_0 = \text{Imbalance} = \text{Accumulation} / \text{Time Interval} \qquad (1c)$$

The imbalance (B_0) is calculated from accumulation over a small time interval (typically 30-60 seconds). The inflow does not have to be measured for B_0 to be calculated. As shown in Fig 2, B_0 represents the total outflow change required to equalize the level, assuming constant inflow in the future. Simply displaying B_0 to the operator after a large disturbance provides valuable information—with this advisory tool, operator can intervene to equalize outflow and inflow with a single change to the outflow, if necessary.

Once the total required change in outflow is known, the 'allowable' accumulation, ΔV_A, (shown by the shaded area in Fig 3a) can be achieved by selecting the appropriate slope of outflow change per unit time. If a straight-line outflow path is assumed (equal moves), then the accumulation is represented by a right triangle with height of B_0 and base of ΔT_E, equalizing time.

The slope of the outflow line can be expressed in terms of B_0 and ΔV_A (Fig. 3b) by substituting for ΔT_E:

$$\text{Slope} = (B_0)^2 / (2 \times \Delta V_A) \qquad (2)$$

The conclusion to be drawn from Eq. 2 is that the allowable accumulation is achieved by setting the slope of the outflow proportional to the square of the imbalance, B_0. See detailed calculation, Figs. 8a, 8b, and 8c.

There should always be a maximum and minimum level limit entered by the operator. The level never exceeds these limits with accumulation control in force. Additional limits on accumulation can be added with equation-based level control, such as having a flexible accumulation limit ('pillow effect') as a function of B_0, e.g:

$$\Delta V_A = \text{Constant} * (B_0)^{0.6} \qquad (3)$$

By comparison, the standard DCS controller uses a P-I algorithm, where the slope of the outflow change, and the accumulation, are both directly proportional to B_0. Therefore, accumulation increases for a P-I controller in direct proportion to the magnitude of the disturbance (Fig 4). If the DCS controller is tuned for a large disturbance, then the outflow may vary excessively during normal operation, causing unsteady feed to downstream units.

Figs 5a and 5b are dynamic simulations of how P-I control responds when the magnitude of the disturbance is doubled. Accumulation in Fig 5b is double the accumulation in Fig 5a (controller tuning is the same in both simulations).

2.2 Create Small Imbalance to Return Level to Set Point within Specified Time

After the outflow and inflow have become equalized, an imbalance must be created in order to return the level to set point (Fig 6a, 6b). The target imbalance (Fig 7a) is calculated from the volume (or mass) differential between set point and current level (ΔV_{SP}) and user-entered time for return to set point (T_R). As the level approaches set point, ΔV_{SP} decreases, and the imbalance is eliminated precisely at the set point (Fig 7b). Even if a disturbance occurs while the level is returning to set point, a new outflow path is calculated such that the imbalance is eliminated precisely at set point (Fig 7c).

3 Drums in Series Applications

Figs 9a, 9b, and 9c show an extension of equation-based level control to drums in series. As soon as the first drum's level begins to change, the future feed to the second drum is known. The outflow path for the second drum can be immediately and correctly calculated (Fig 9b).

This type of feed-forward control, based on the upstream imbalance rather than feed flow, increases the time available for the second drum to manage the disturbance, resulting in a more gradual shift in outflow from one steady state to the next.

3.1 Ethylene Plant Application: Demethanizer Feed System

Fig 10 is a schematic of a demethanizer column with its associated feed drums. Reactors and distillation systems downstream require steady feed for best operation. Large feed rate disturbances occur in ethylene plants whenever furnaces are taken offline for decoking or placed into service after decoking.

In this application of drums in series, the tower receives liquid from each of the 3 feed drums. Methane and hydrogen in the 3 feeds are recovered in the distillate, while ethylene and heavier are recovered as bottom product. A 'recovery factor' from each feed drum represents the ethylene and heavier percentage of each feed stream, estimated from the design material balance. Therefore, the bottoms recovery factors (R_1, R_2, and R_3) of ethylene and heavier can be used to predict the column bottom's future required liquid outflow rate.

Changes to the column bottoms outflow, along the calculated path, begin immediately when the levels of the feed drums begin to change. Not only does this spread the outflow change over a longer time period, but bottoms accumulation can be more tightly limited, maintaining nearly constant flow through the thermosiphon reboiler during the entire event.

3.2 Distillation Application: Feed Drum in Series with Reflux Drum

A more general application of drums-in-series control can be used on any distillation system with a level-controlled feed drum (Fig 11). The feed drum combines with the reflux drum and/or bottoms level control to further dampen feed disturbances while greatly reducing the likelihood of overfill or emptying.

With drums-in-series control it becomes possible to safely increase the set points on feed and product level controllers. This increases residence times, thereby dampening composition variations in feed and product streams. Increasing the residence time of the reflux drum also reduces variation of the reflux, which represents a huge benefit for the entire system.

3.3 Dead Time Application: Distillation System Designed with No Reflux Drum

Successful application of equation-based level control requires proper accounting of dead time between flow valve and level measurement. Most level control applications have minimal dead time, but there are examples where dead time affects performance.

One example involves a C3 splitter application with no reflux drum (Fig 12). Since the distillate flow is the larger of the two product streams, it must be used for inventory control of the system, which is measured as the level in the bottom of the tower. Several minutes of dead time exist between the distillate flow and bottoms level.

The user creates an array of distillate flow changes that have occurred during the dead time period. The array is updated on every execution of the algorithm. The calculated imbalance is adjusted by the algebraic total of outflow moves in the array, thereby ensuring correctly calculated distillate flow moves.

By comparison, a P-I controller in this application would have to be significantly de-tuned in order to avoid oscillatory behavior.

3.4 Vapor Phase Reactor Application: Partial Pressure and Total Pressure Control

Fig 13 is a schematic of a vapor-phase reactor with 3 reactants. Nitrogen is also fed as it conveys solid catalyst to the reactor. The objective is to maintain constant product purity by controlling the partial pressure of each reactant, as well as total reactor pressure. Partial pressure can be difficult to control directly, because a change in any reactant flow rate, or vent flow, changes partial pressure of all reactants. Steady state can be difficult to achieve.

Instead, the apparent 4x4 multivariable control problem can be handled as 4 separate inventory control problems, one for each reactant, and one for total pressure. Since the total vapor volume and temperature of the reactor are known, the partial pressure each reactant can be calculated at any time from the measured vent composition and total pressure (Eq. 3a).

$$P_i = Y_i \text{ x } P_{total} \qquad (3a)$$

where Pi and Yi refer to partial pressure and mole fraction of each reactant in the vapor space.

Similarly, the target partial pressure of each reactant can be calculated by substituting target pressure and composition for measured values above.

$$P_i \text{ (targ)} = Y_i \text{ (targ) x } P_{targ} \qquad (3b)$$

The change in reactant partial pressure over a short interval is calculated in Eq. 3c, and the change in mass inventory of that reactant (B_0) is calculated in Eq 4.

$$\Delta P_i = \Delta (Y_i \text{ x } P_{total}) \quad (3c)$$

$$B_0\ (i) = \Delta\ \text{Mass}\ (i)\ /\ \text{time} = MW_i \times \Delta P_i \times V\ /\ (Z \times R \times T)\ /\ \text{time} \quad (4)$$

where MWi represents the molecular weight of each reactant. In order to control the reactant inventory, we calculate the difference between current and target mass for each reactant, Eq 5.

$$\Delta\ \text{Mass to target}\ (i) = MW_i \times (P_i - P_i\ (\text{targ})) \times V\ /\ (Z \times R \times T) \quad (5)$$

This is enough information to control the reactant inventories by manipulating the feed flow of each, using the methods described above for liquid levels. If feed composition, reaction rate, or vent rate changes, these changes are reflected in the measured composition of the vent, so that target inventory of each reactant is achieved without going beyond.

Many vapor phase reactors, including polymerization reactors, can make a variety of different products. Each product has its own requirements for vapor composition and total pressure. Reactor operation can be switched from one product to another, to satisfy business needs. Vapor composition and total pressure must be shifted to the new targets as quickly as possible, and without overshoot. Product that does not conform to the new requirements may have to be reprocessed or sold as off-spec material.

When the target inventories change, Eq. 5 can be used to independently bring all reactants to their new targets, within the desired time, using the 'return to set point' calculations for liquid levels.

4 Conclusion

Equation-based level control has been discussed in detail, and compared with traditional P-I control. The advantages of equation-based control include the use of equal outflow moves to directly control accumulation, and return to set point without oscillation.

Applications of equation-based level control, including demethanizer feed system, distillation systems with a feed drum, systems with dead time, and vapor phase reactors have been discussed.

Enough detailed information has been given in this paper for process engineers with general process control knowledge to write and implement their own level control algorithms.

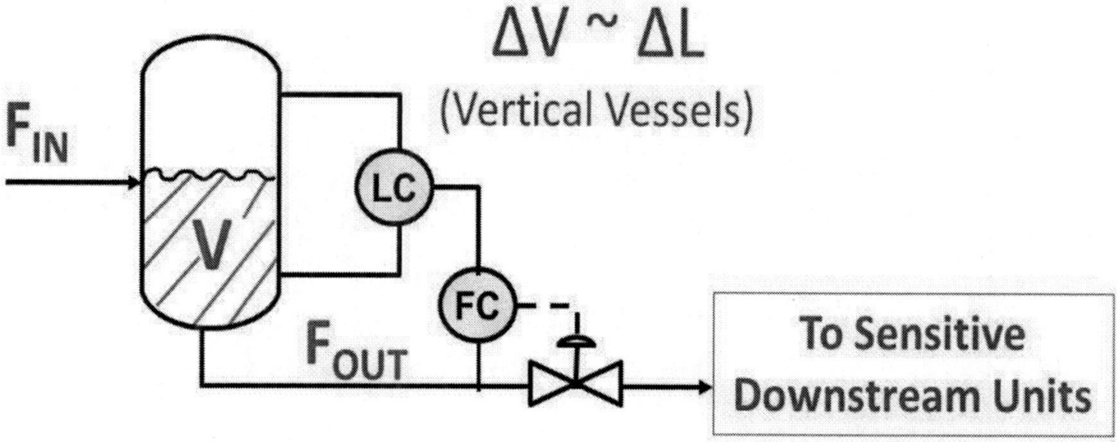

Fig 1: Typical DCS Level Control Arrangement

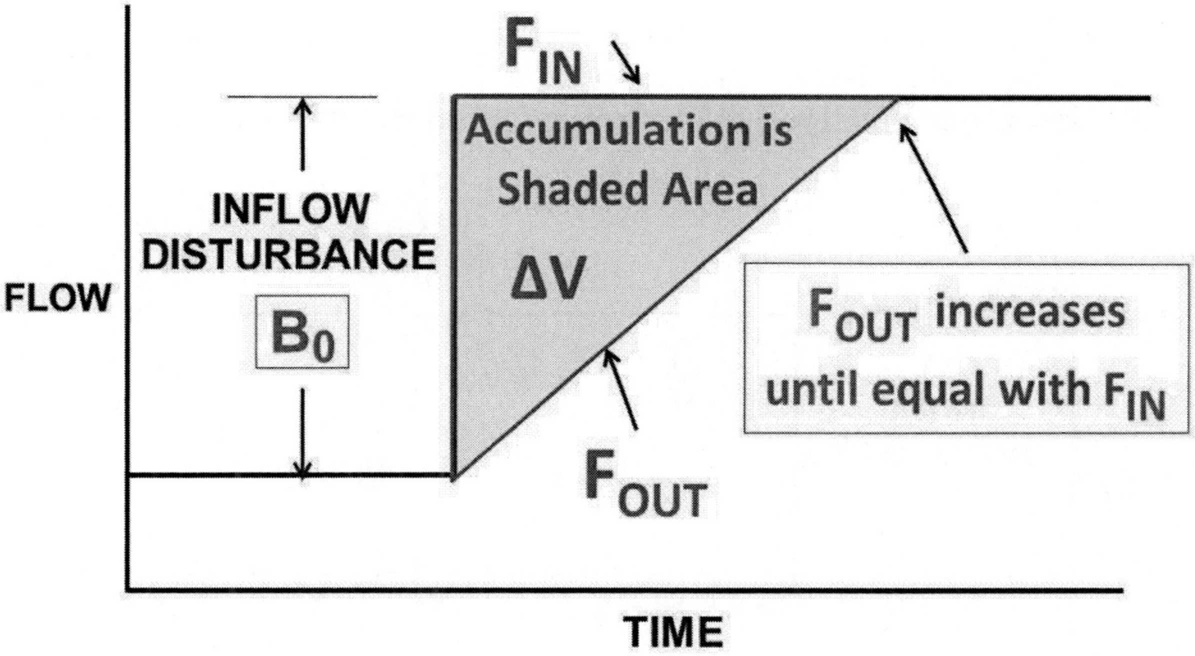

Fig 2: Inflow, outflow, and accumulation

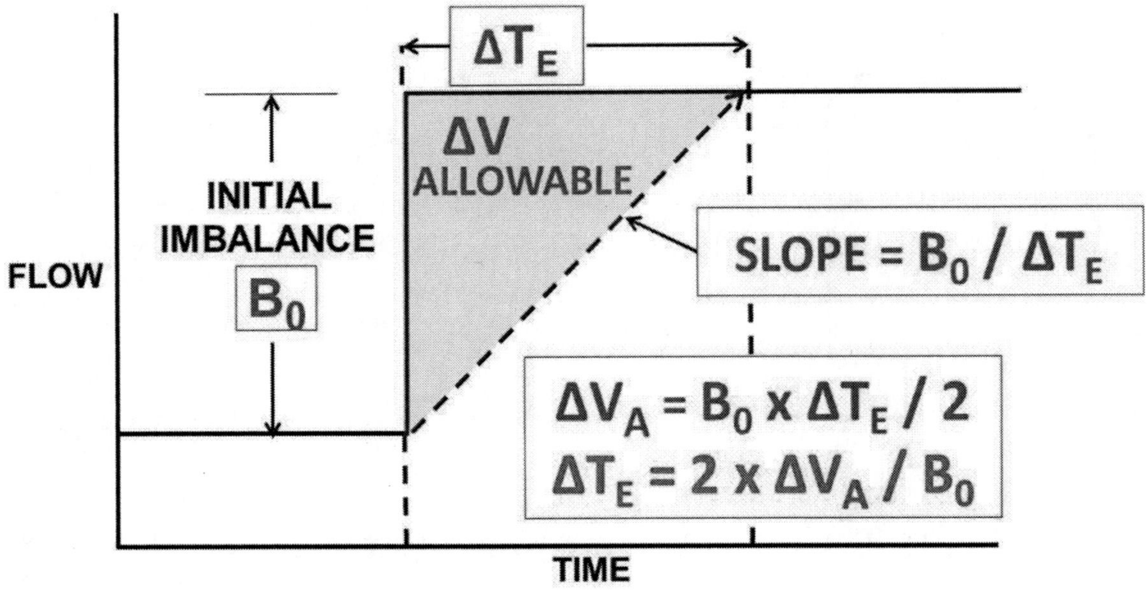

Fig 3a: Calculate slope of outflow required to limit accumulation

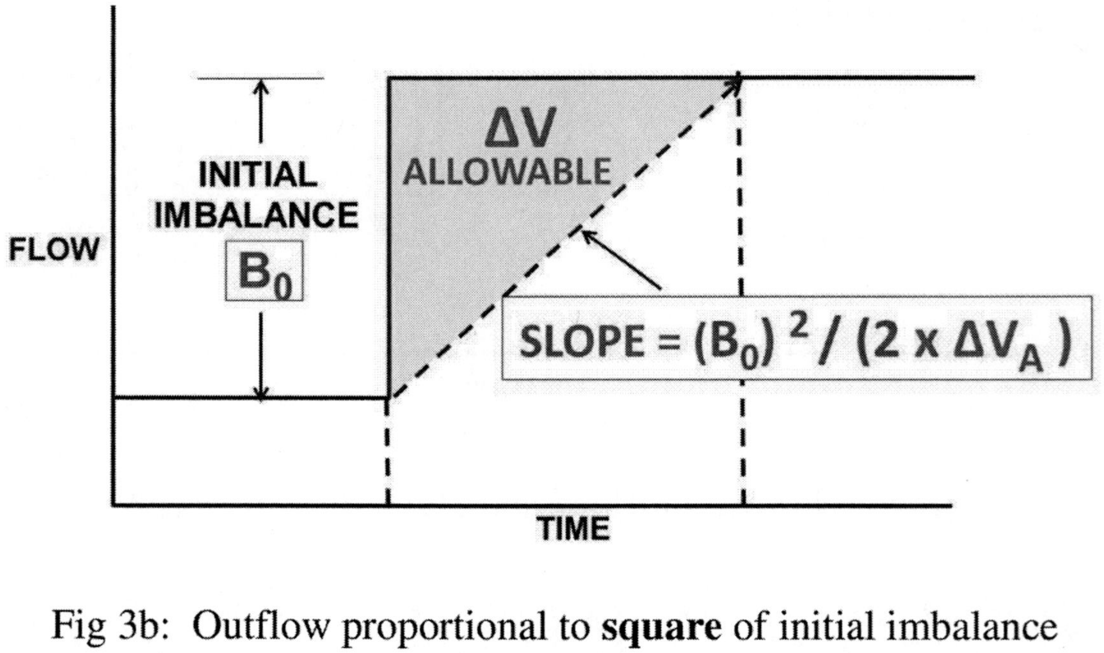

Fig 3b: Outflow proportional to **square** of initial imbalance

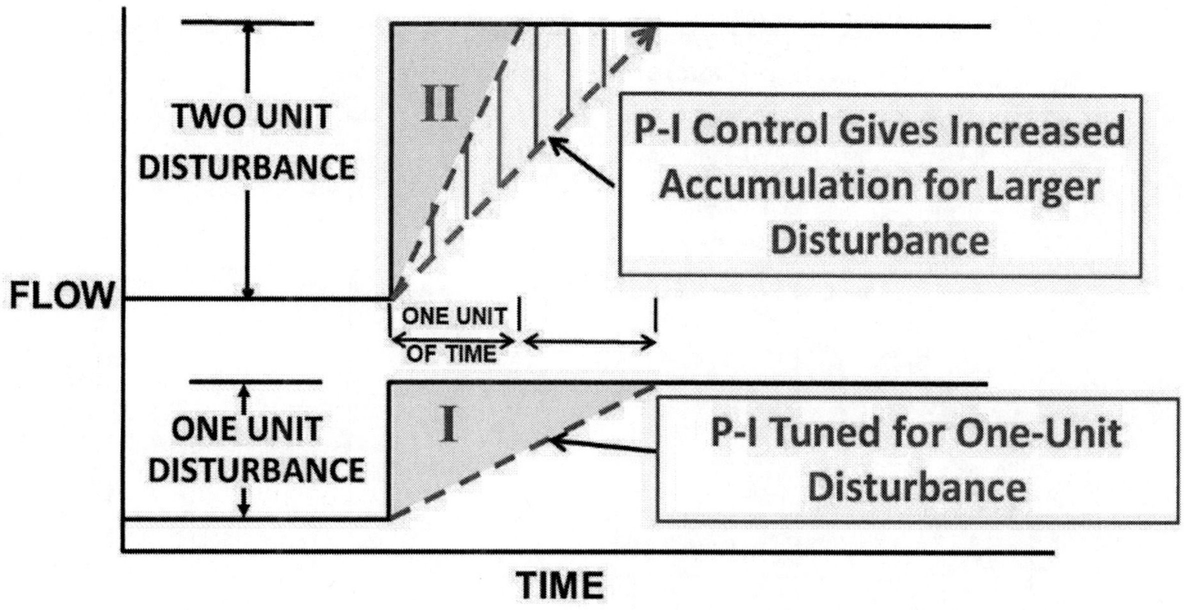

Fig 4: P-I control gives accumulation proportional to disturbance

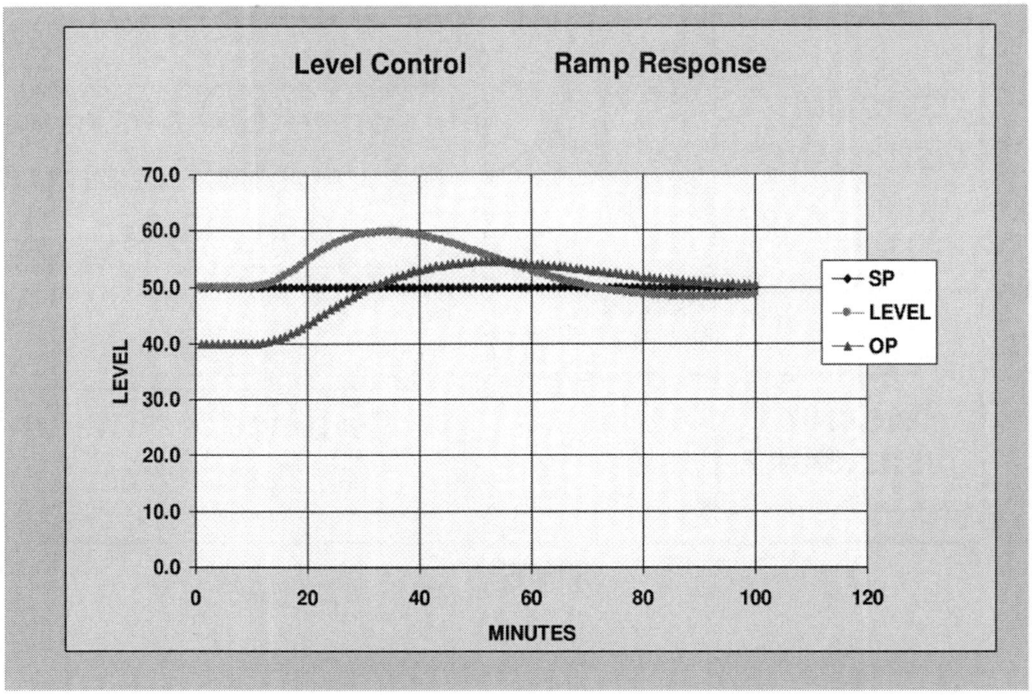

Fig. 5a: Simulated P-I Level Control on One-Unit Disturbance

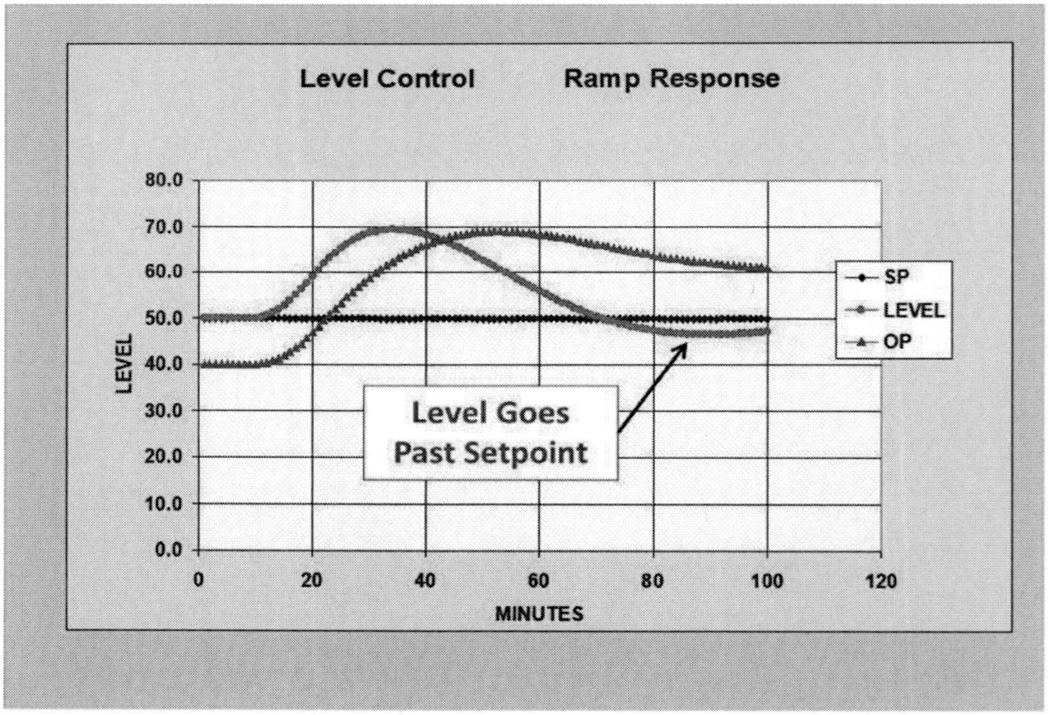

Fig. 5b: Simulated P-I Level Control on Two-Unit Disturbance

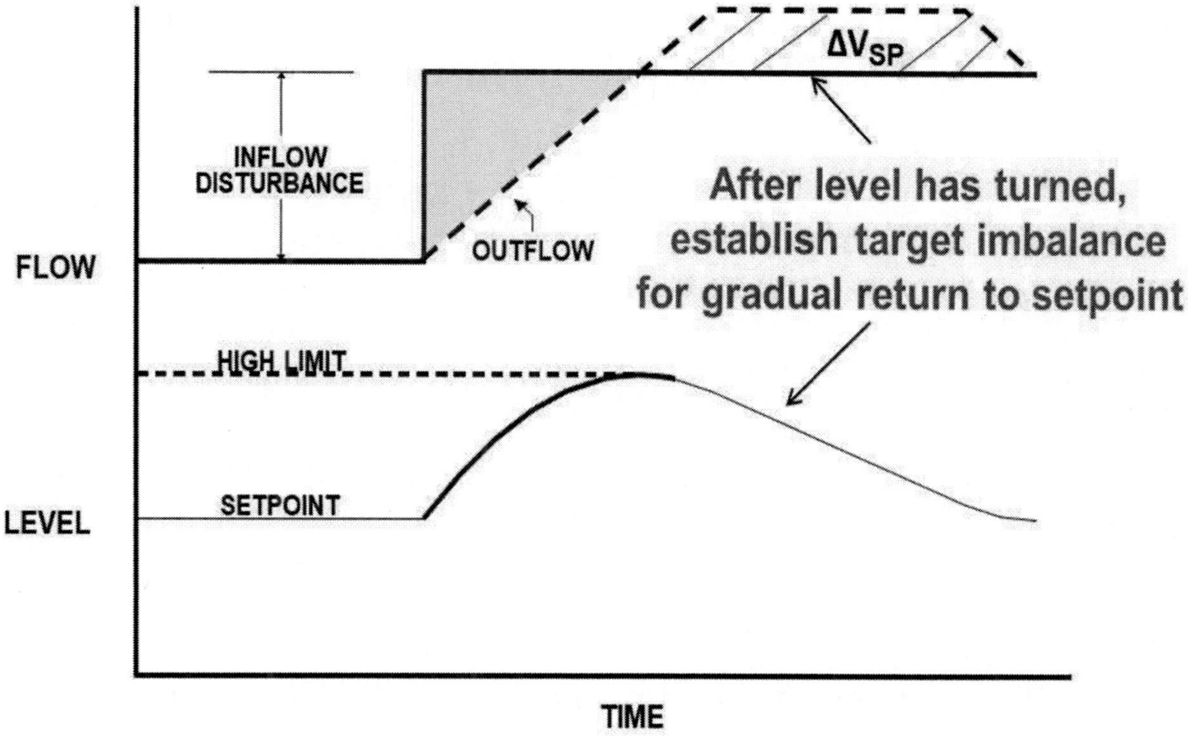

Fig 6a: Target imbalance is established for return to setpoint

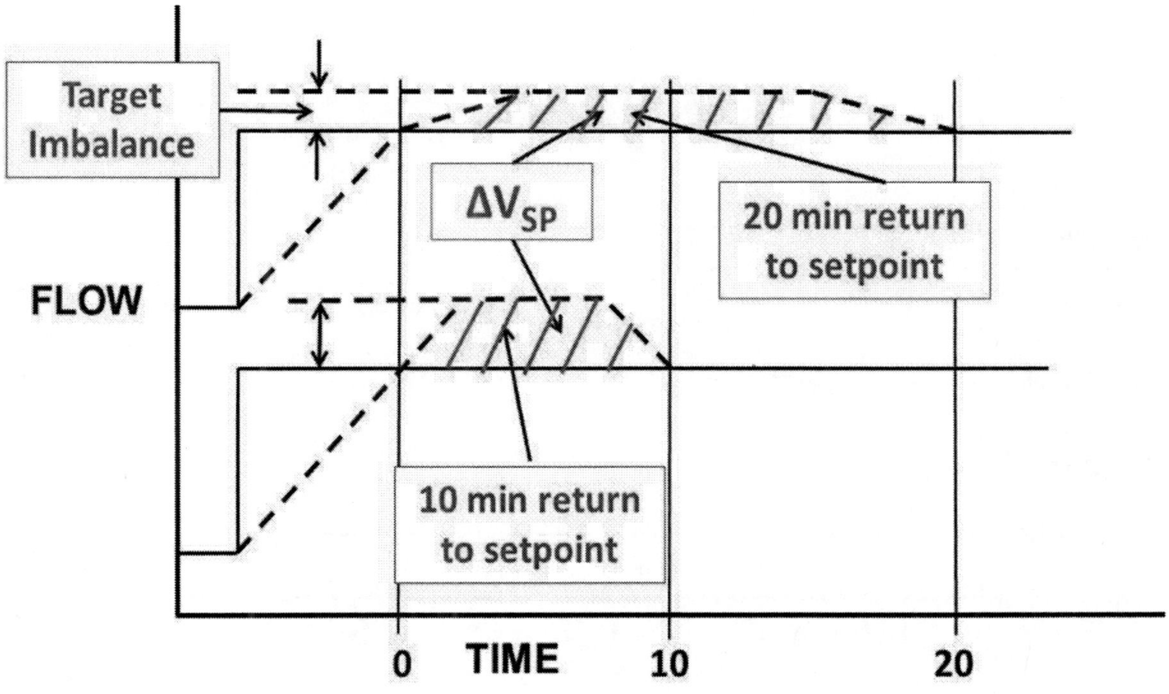

Fig 6b: Time for return to setpoint is a basic concept

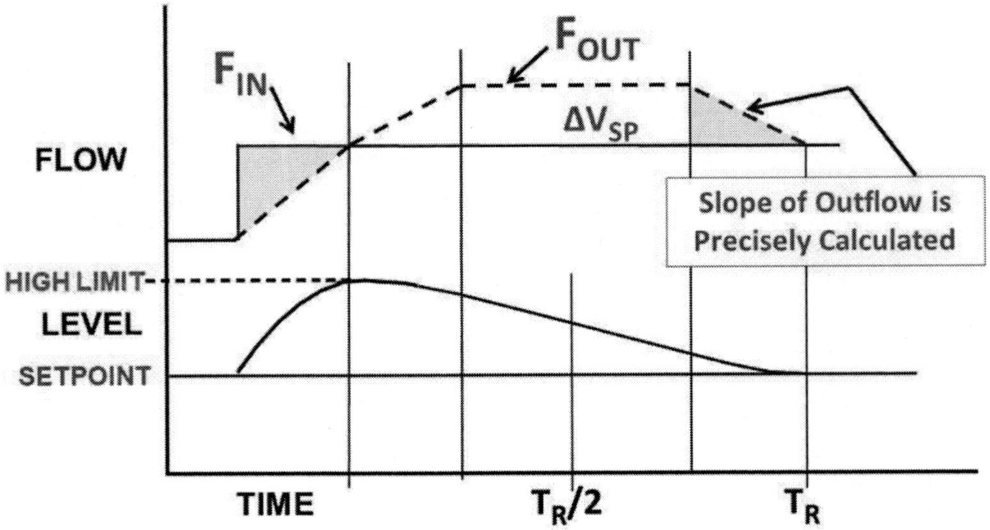

Fig 7a: Imbalance is eliminated precisely at setpoint

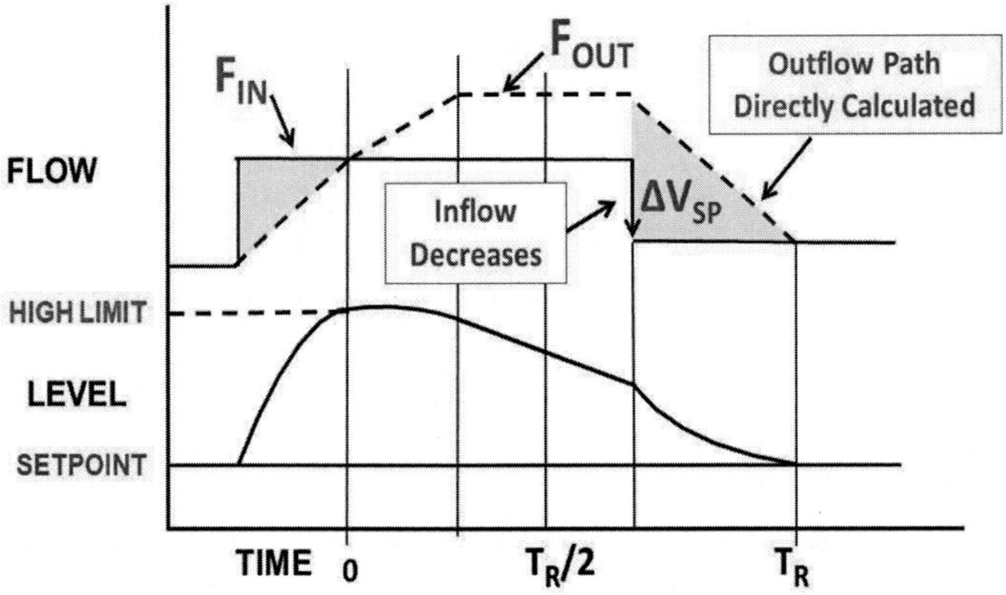

Fig 7b: Large disturbance on return to setpoint is handled correctly

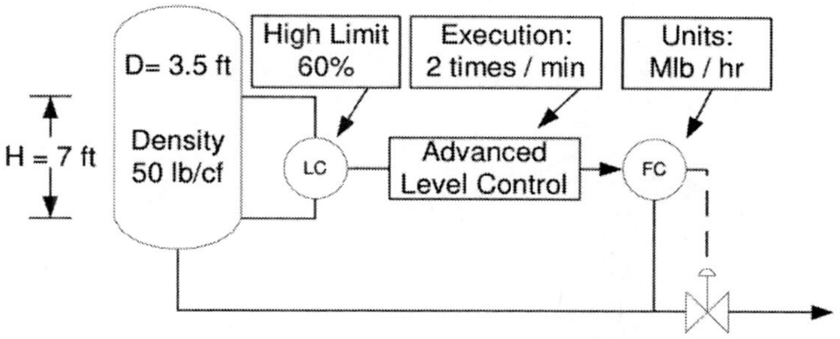

Time, sec	Level, %	FC SP	Liq Vol, cf	Liq Mass lb
T_0	52.0	20,000	35.02	1751
$T_0 + 30$	52.3	?	35.22	1761

Fig. 8a: Example Calculation – Convert Level Change to Mass

Imbalance = (1761 - 1751) lb / 30 sec = 1200 lb / hr
Ultimate flow target = 20,000 + 1200 = 21,200 lb / hr

Liq Mass @ 60% = 40.41 cf x 50 lb / cf = 2020 lb
Accumulation = 2020 - 1761 = 259 lb

Flow move = [1200 x 1200 / (2 x 259)] / 120 exec / hr
= 23.2 lb / hr / execution

Time to equalize = 1200 / 23.2 = 51.7 exec = 25.9 min

Fig. 8b: Imbalance, Outflow, and Equalizing Time Calculated

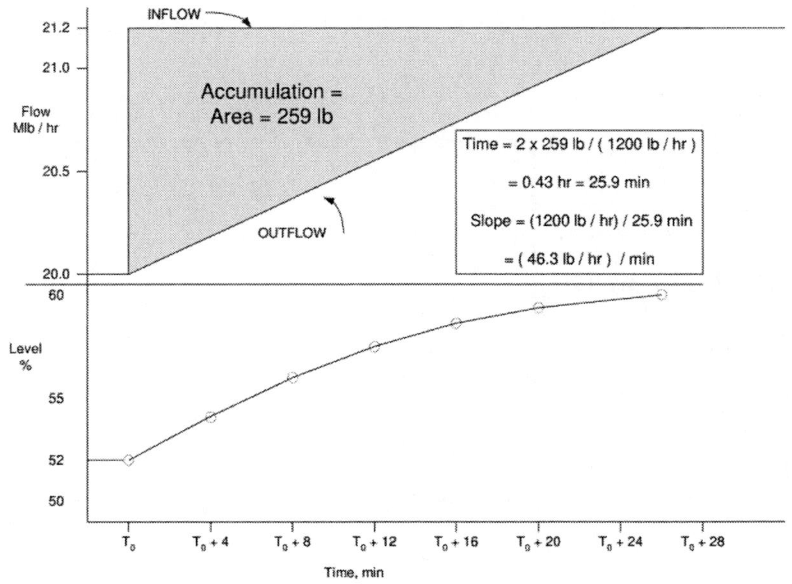

Fig. 8c: Graphic Representation of Outflow Calculations

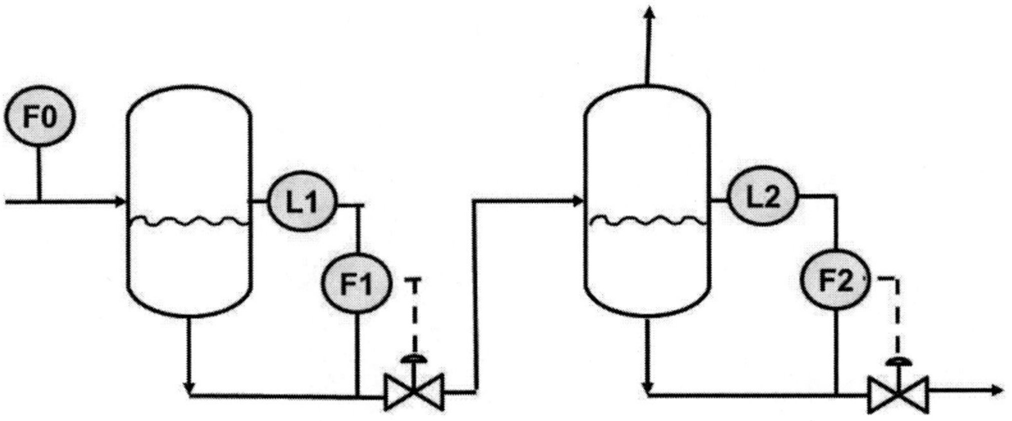

Fig. 9a: Increased Outflow from Drum #1 Goes to Drum #2

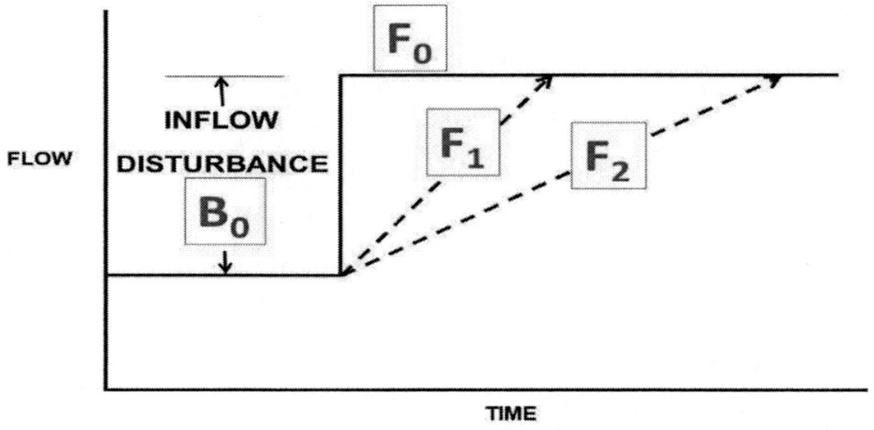

Fig. 9b: Outflow Path from Second Drum Directly Calculated

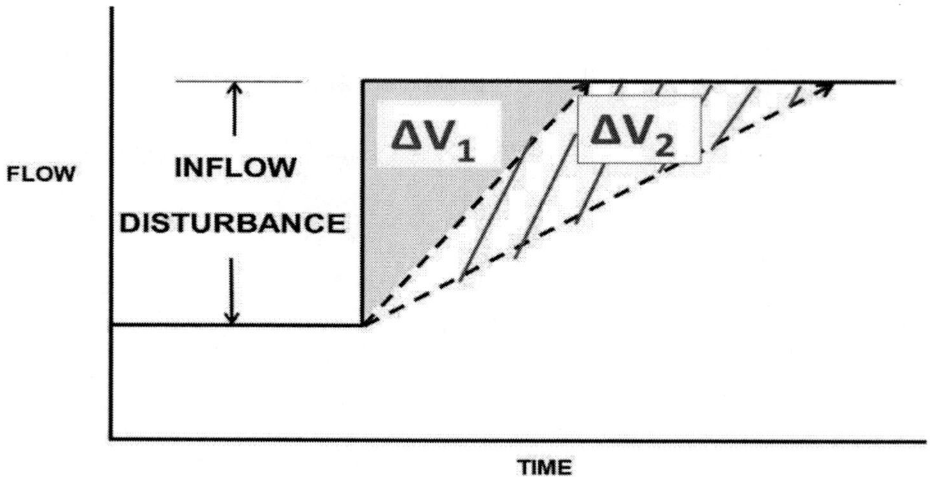

Fig. 9c: Imbalances and Volumes are Additive

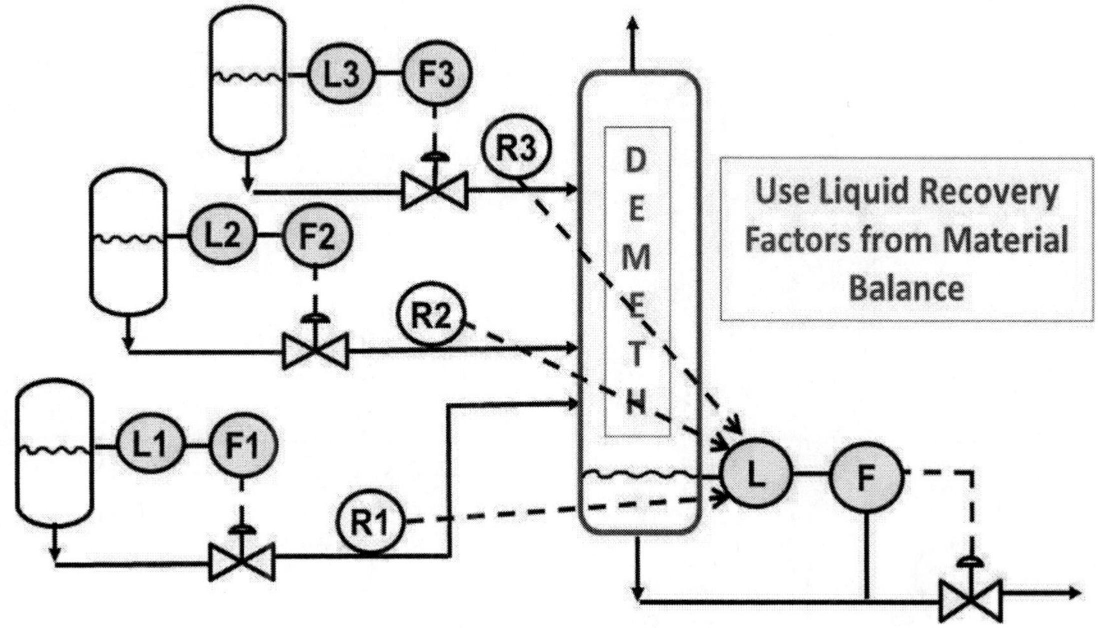

Fig 10: Ethylene Plant Application of Drums-in-Series

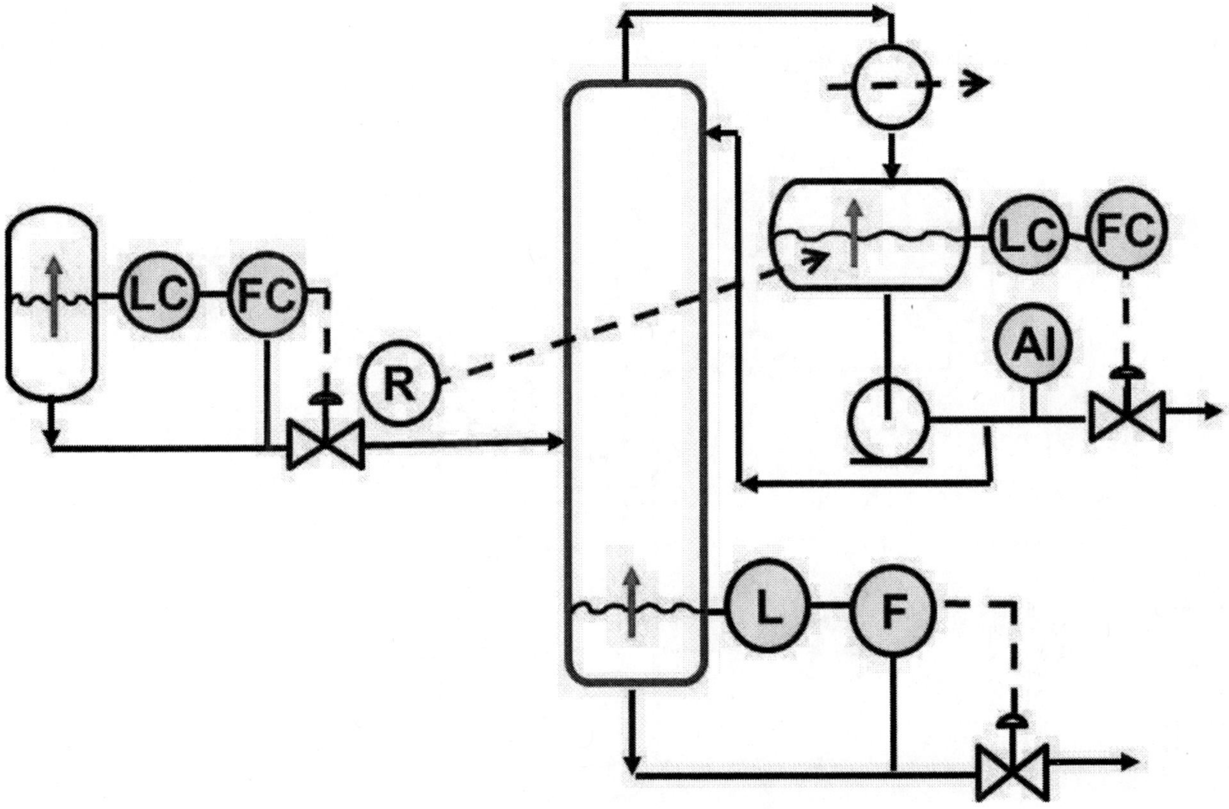

Fig. 11: Increase Level Set Points to Dampen Composition Variations

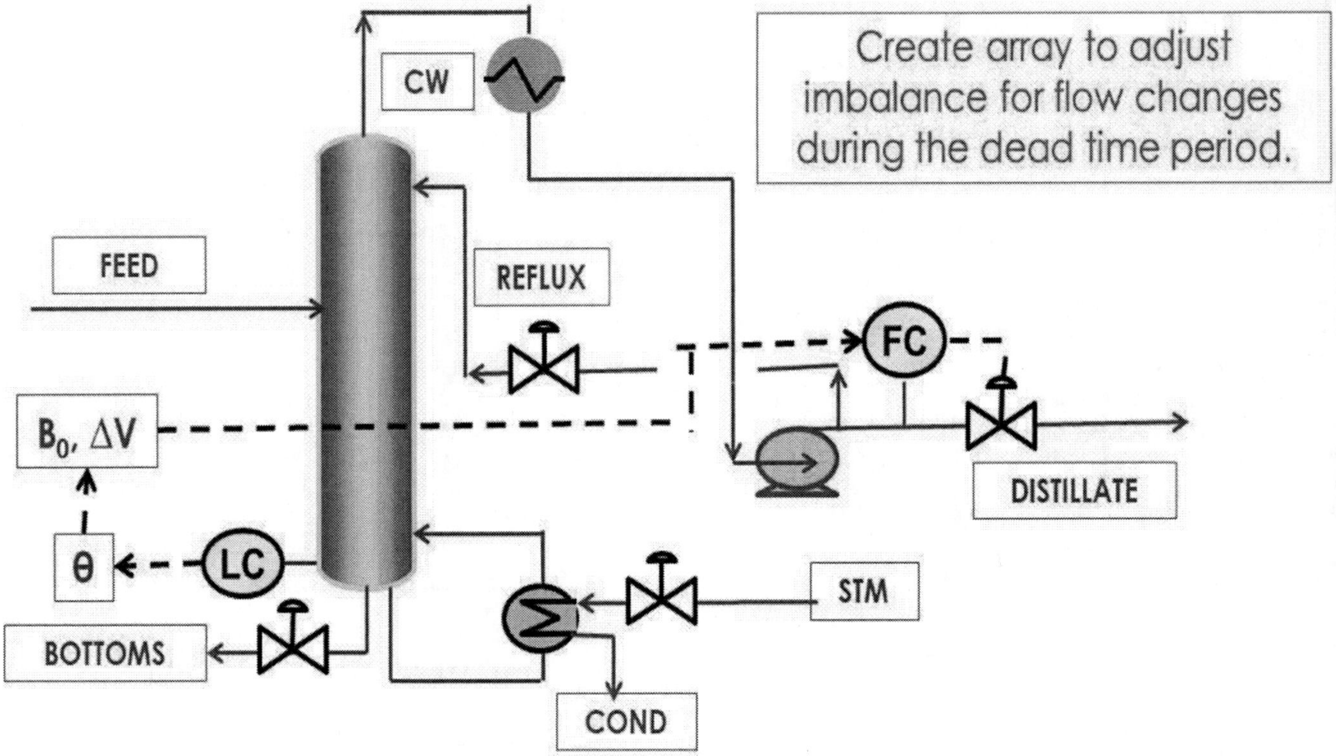

Fig 12: Create array to adjust for outflow moves during dead time period

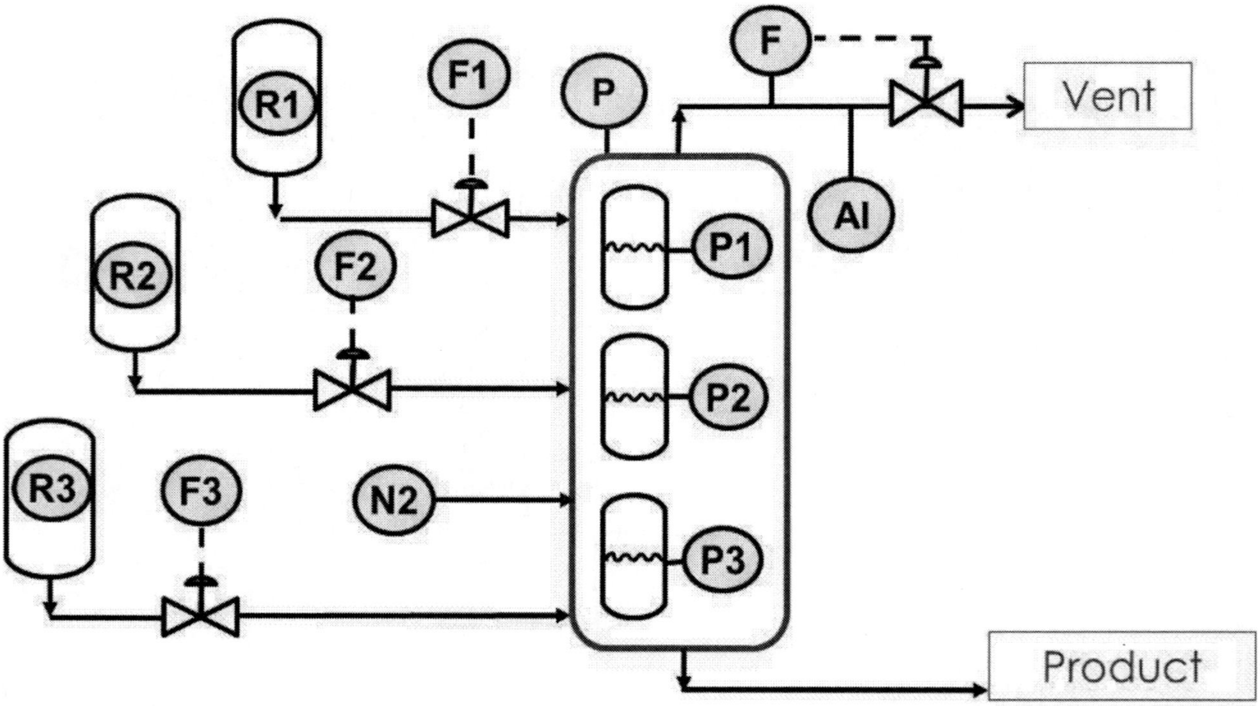

Fig 13: Component Material Balances Maintain Reactor Composition

96c Advancing Petrochemical Processes Inferential Analysis with Eigen's ML Transfomers

Erdin Guma

Eigen Control Inc. is leveraging the power of Transformer models in Inferential Analysis of Product Properties in distillation and reactor processes. Transformer models are a cutting-edge development in artificial intelligence (AI). We believe it represents a significant shift from traditional empirical based inferential methods which enhances efficiency, scalability, and precision of product quality inferential predictions.

Historically, the concept of Transformers gained momentum with the seminal 2017 research paper 'Attention Is All You Need' by the Google Brain team. This paper was instrumental in revolutionizing sequence-to-sequence learning, setting a new precedent in the AI realm. Subsequent adaptations and advancements, such as the 2018 GPT framework, emphasized the self-attention mechanism's role, scalability, and parallelization—each contributing to the complex architectural evolutions and industry agnostic practical applications seen today, such as that of ChatGPT from OpenAI.

The application of Transformer models in petrochemical processing control presents a paradigm shift in handling sequence-to-sequence data and understanding intricate process dependencies. These models, recognized for their self-attention mechanisms, make them particularly suited for the time-sensitive and data-rich environment of refinery operations. The efficiency of Transformers stems from their parallel architecture, allowing real-time processing even on standard low-power computers commonly used in plant control networks. Moreover, their scalability is demonstrated in their adeptness at managing growing data volumes and model complexities, essential for adapting to highly nonlinear processes inherent in refinery control.

This presentation will explore the application and customization of Transformer technology to separation processes, particularly as it applies to a Crude Unit and an NGL Fractionation column. A significant challenge lies in merging Transformer architecture with chemical engineering expertise, encompassing a) equipment and instrument design, b) process design, and c) operations. Additionally, we face the hurdle of incorporating classical mathematics and control theory, such as Kalman filters, for data validation and integrity in real-time applications. We will also discuss the challenges and opportunities presented by the software infrastructure supporting Transformer technology in a control environment. This includes the development and integration pipeline crucial for enhancing the resilience, security, and availability of Transformer-based inferential applications for closed-loop control.

In conclusion, we will present the outcomes of an actual deployment in an NGL Fractionation plant, highlighting the superior accuracy of Transformer-based model results as a significant advantage of this emerging technology.

172a Unconventional Feedstock Processing

Hsiang Yee Hoekstra, Eloy Flores III, Veshal Venkat

At SwRI, we engage in the process development and improvement for a variety of sustainable fuels and chemicals from lab to pilot scales. The applied research, development, testing and evaluation methods undertaken help bridge the gap between fundamental science and commercialization. The technologies developed include but are not limited to hydrogen production, CCUS, renewable fuels, chemicals and materials production and conventional thermos, photo and electro-chemical conversions. One such technology is the High Pressure Circulating Fluidized bed (HPCFB) reactors. HPCFB reactors have recently showcased itself as a promising technology for sustainable fuel and chemical conversions. These reactors have several advantages over conventional reactors, including higher rates of heat and mass transfer, improved mixing of reactants, and the ability to recycle catalysts with high recirculating solid flow rates for heat recovery. HPCFBs are particularly suitable for high-temperature, high-pressure processes such as gasification, pyrolysis, and hydrotreating. At SwRI, we have developed HPCFBs and the supporting infrastructure for practical applications, ranging from small-scale laboratory experiments to larger pilot-scale CFBs. These reactors can efficiently convert both renewable and petrochemical feedstocks into valuable end products. The reactor system has also shown promise for the one-step cracking, deoxygenation, and hydro-isomerization of biogenic feedstocks into specific fuels and chemicals.

SPRING24
+20TH GCPS
A Joint AIChE and CCPS Meeting

Influence of Feedstock Variability on Corrosion Dynamics in Municipal Waste-to-Biofuel Conversion Process

Russ Davis
Mistras Group
4000 Underwood
LaPorte, TX 77571
russ.davis@mistrasgroup.com

Sudhakar Mahajanam, PhD, MBA, PE
Mistras Group
4000 Underwood
LaPorte, TX 77571
sudhakar.mahajanam@mistrasgroup.com

Prepared for Presentation at
American Institute of Chemical Engineers
2024 Spring Meeting and 20th Global Congress on Process Safety
New Orleans, LA
March 24-28, 2024

Influence of Feedstock Variability on Corrosion Dynamics in Municipal Waste-to-Biofuel Conversion Process

Russ Davis, Sudhakar Mahajanam

Abstract

Municipal waste contains a plethora of organic and inorganic chemicals. While the front-end separation and chomping process removes many of the inorganic materials, many of these inorganic compounds still get processed through the gasifier. The process stream from the gasifier contains sulfur compounds, organic and inorganic acids, halides and other corrodents that negatively affect the equipment.

Materials of construction are selected by the feed engineering group based on a given assumption of the process constituents in the feedstock. The materials are then specified based on anticipated damage mechanisms for a typical service life of 25 to 30 years. When feed streams change, the assumptions that were used to select materials are no longer valid. The problem with municipal waste feed stock is that the chemical composition is unpredictable and anticipated damage mechanisms change throughout the process. This can lead to higher-than-expected corrosion rates and catastrophic cracking phenomena, causing premature failure.

Asset integrity and reliability programs must be very robust. Every potential damage mechanism must be identified with the understanding that feed stock will change. Inspection and testing must be performed to ensure damage is detected long before a loss of containment event. Feed stream analysis must be evaluated daily in order to predict downstream damage.

The benefits of using municipal waste to meet America's chemical and fuel needs are great. Preventing loss of containment events and keeping these plants safe is a new asset integrity challenge that must be met.

Keywords: Corrosion, Cracking, Biofuel, Municipal Waste, Feedstock.

Introduction

This paper aims to address the unique challenges associated with anticipating damages in plants that break down sorted organic municipal waste into synthetic gas (syngas) to produce biofuels. The focus is on the mechanical integrity aspects in the context of biofuel production facilities, specifically those producing kerosene, jet fuel, and other biofuels. The paper discusses the key factors that contribute to equipment degradation, potential failure modes, and the implementation of proactive strategies for damage prevention and control.

The conversion of sorted organic municipal waste into syngas as a precursor for biofuel production presents both opportunities and challenges. These facilities are at the forefront of sustainable energy, but they also face unique operational and mechanical integrity issues. The primary concern is the anticipation and management of damages that can affect the efficiency, safety, and longevity of the plant.

1. Gasification Process

In the context of converting sorted organic municipal waste into syngas for biofuel production, the gasification process plays a pivotal role.[1,2] This process, predominantly comprising the stages of pyrolysis, oxidation, and reduction, involves a series of complex thermochemical reactions. Understanding these stages in detail is crucial for optimizing the gasification process for efficient syngas production. This section of the technical paper expands on each of these stages, delving into the intricate chemistry, temperature profiles, and relevant equipment involved.

1.1 Pyrolysis: The Onset of Thermochemical Decomposition

The first critical stage in the gasification process is pyrolysis, where organic material undergoes thermal decomposition in the absence of oxygen. This stage sets the foundation for subsequent reactions in the gasification process.

1.1.1 Temperature Range and Significance: Occurring at temperatures between 350°C and 700°C, pyrolysis initiates the breakdown of complex organic molecules into simpler compounds. This temperature range is carefully maintained to optimize the yield of pyrolysis products.

1.1.2 Chemical Transformations: In this stage, long-chain hydrocarbons are broken down into shorter molecules. The process results in the production of char, a carbonaceous residue; tar, a mixture of various organic compounds; and pyrolysis gas, primarily composed of CO, H_2, and small hydrocarbons.

1.1.3 Equipment Used: The pyrolysis reaction is typically carried out in reactors such as rotary kilns or fluidized bed reactors. These reactors are designed to provide uniform heat distribution and accommodate the variable nature of municipal waste.

1.2 Oxidation: Generating the Necessary Heat

Following pyrolysis, the partially decomposed material enters the oxidation stage, where it is partially oxidized.

1.2.1 Temperature Control: This stage takes place at higher temperatures, ranging from 800°C to 1,000°C. The elevation in temperature is critical for sustaining the gasification reactions in the subsequent reduction stage.

1.2.2 Oxidative Reactions: A controlled amount of oxygen, or air, is introduced to react with the pyrolysis products. This controlled oxidation is exothermic, releasing substantial heat which is essential for the endothermic reactions in the reduction stage.

1.2.3 Equipment and Design Considerations: Oxidation chambers are engineered to ensure optimal mixing of oxygen and pyrolysis products. The control of oxygen flow and the chamber's thermal insulation are key to maintaining the desired temperature and reaction efficiency.

1.3 Reduction: The Heart of Gasification

The reduction stage is where the core gasification reactions occur, transforming the output of the oxidation stage into syngas.

1.3.1 High-Temperature Environment: Occurring at temperatures ranging from 800°C to 1,400°C, the reduction stage is where the bulk of syngas is produced. The high temperature is crucial for driving the endothermic reactions involved.

1.4 Key Chemical Reactions:

1.4.1 Water-Gas Shift Reaction: Char reacts with steam (H_2O) to produce carbon monoxide (CO) and hydrogen (H_2). This reaction is fundamental in increasing the hydrogen content of the syngas.

1.4.2 Boudouard Reaction: Char also reacts with carbon dioxide (CO_2) to form additional CO. This reaction helps in adjusting the CO:H_2 ratio in the syngas.

1.4.3 Methanation: Additionally, some CO and H_2 may react to form methane (CH_4), contributing to the energy content of the syngas.

1.5 Gasifier Technologies: The reduction stage is facilitated in gasifiers, which are categorized as fixed bed, fluidized bed, or entrained flow gasifiers. The choice of gasifier is influenced by factors like the nature of the feedstock, desired syngas composition, and process efficiency.

2. Overview of Corrosion Mechanisms

In plants converting organic municipal waste to syngas for biofuel production, equipment is often exposed to a variety of chemical impurities. These impurities can induce several specific corrosion mechanisms, affecting the integrity and lifespan of the plant's equipment. Understanding these mechanisms is critical for selecting appropriate materials and implementing effective maintenance strategies.

2.1 Specific Damage Mechanisms Induced by Chemical Impurities

2.1.1 Acidic Corrosion

2.1.2 Impurities Involved: Sulfur compounds, chlorides, low pH organic acids.

2.1.3 Mechanism: These compounds, often present in municipal waste, can lead to the formation of acids during processing. Acidic environments accelerate corrosion, especially in carbon steel and certain stainless steels.

2.1.4 Preventive Measures: Utilization of acid-resistant alloys, such as high-grade stainless steels (e.g., SS 316), and application of protective coatings.

2.1.5 Chloride-Induced Stress Corrosion Cracking (SCC) Impurities Involved: Chlorides, often from kitchen waste or certain industrial waste components.

2.1.5.1 Mechanism: Chlorides can induce cracking in stainless steels, particularly at elevated temperatures. This is a concern in areas like heat exchangers and boilers.

2.1.5.2 Preventive Measures: Use of duplex stainless steel or super austenitic stainless steels, which have higher resistance to chloride-induced SCC.

2.1.6 Sulfidation

2.1.6.1 Impurities Involved: Sulfur compounds, commonly found in rubber or certain food wastes.

2.1.6.2 Mechanism: High temperatures can cause sulfur compounds to react with metal, leading to sulfidation, which weakens the metal structure.

2.1.6.3 Preventive Measures: Implementation of low-alloy steels with added chromium and molybdenum, which are more resistant to sulfidation.

2.1.7 Hydrogen Embrittlement

2.1.7.1 Impurities Involved: Hydrogen, which can be a by-product in certain waste processing stages.

2.1.7.2 Mechanism: Hydrogen atoms can diffuse into metal, leading to cracking and embrittlement, particularly under high-pressure conditions.

2.1.7.3 Preventive Measures: Use of low-strength, hydrogen-resistant materials and proper heat treatment methods to reduce susceptibility.

2.1.8 Galvanic Corrosion

2.1.8.2 Impurities Involved: Dissimilar metals in the presence of an electrolyte.

2.1.8.2 Mechanism: When two different metals are in contact, the more anodic one corrodes faster than it would alone, while the cathodic metal corrodes slower.

2.1.8.3 Preventive Measures: Careful selection of compatible materials, use of insulating gaskets or coatings to prevent direct metal-to-metal contact.

2.1.9 Erosion-Corrosion

2.1.9.1 Impurities Involved: Solid particles in the waste stream, such as grit or glass.

2.9.2 Mechanism: Mechanical wear exacerbated by corrosive elements, common in pumps and piping.

2.1.9.3 Preventive Measures: Use of erosion-resistant materials like high-alloy stainless steel, and design modifications to reduce flow turbulence.

3. Materials and Corrosion Considerations

A literature search was conducted to get a better understanding of the materials typically used in these waste-to-energy processes, as well as the predominant damage mechanisms observed.

High temperature superheater corrosion was identified as the key factor limiting the efficient use of municipal waste fired boilers in the early 2000s and setting a maximum temperature threshold of 450°C.[3,4] Biofuels are renewable, which means they do not add CO_2 to the environment, unlike traditional fuel sources like coal and oil. However, they are volatile, have a high hydrogen content, and contain sulfur and chlorine. Two forms of corrosion were found to be dominant: alkali chloride corrosion from sulfation of fly ash/fuel deposits (also termed as "chlorine circulation theory", "memory effect" or "autocatalytic corrosion"), and molten phase corrosion. In addition, fouling and high temperature oxidation were also reported. Field experiments indicated that a S/Cl ratio greater than 2.0 was desired to minimize corrosion leading to recommendations that co-combustion of these biofuels with high sulfur fuels such as oil, peat or coal be performed to enhance efficiencies.[3,5] It was also found that higher amount of deposited chlorides lowered the melting point of the ash deposits (known as the "first melting temperature"), thereby increasing the molten phase.[4] Various candidate alloys were tested in lab and field trials for use as superheater tubing materials: 410 SS, 347 SS, X20 Fe-Cr alloy, Alloy 625 and 310HCbN SS.[3,4] The high Cr-No Alloy 625 performed the best in highly corrosive conditions in part due to the presence of molybdenum (Mo) as an alloying element as well as the protective chromium oxide (Cr_2O_3).[4] More recent studies have included a wider range of alloys: carbon steel, ferritic steels (T11, T22), 304H SS, Alloy 28, Alloy 800H in addition to the above alloys.[5] It was reported that the protective nature of the Cr_2O_3 layer was lost due to fluxing with alkali chlorides to form chromates. Another form of oxide breakdown occurred due to volatilization of Cr_2O_3 in the presence of high temperatures, humid atmospheres and rapid gas flow.

Alloy 625M and C-276M weld overlay tubing and supersonic HVOF flame sprayed coatings (with NiCrSiB alloy) have been successfully tested for corrosion protection of water wall regions of the boilers.[4] A more recent advancement utilized a cheaper, leaner Alloy 31 overlay instead of Alloy 625 in the less aggressive upper furnace regions of the boiler with no loss of corrosion protection.[6] A high Cr-Ni thermal spray applied (TSA) HVOF coating (Ni-57CrMoBSi alloy) displayed good erosion-corrosion resistance in a CFB combustor even with chlorides present.[5] A comprehensive TWI project aimed at mitigating chlorine-induced corrosion, solid particle erosion and ash build-up evaluated several TSA formulations and ceramic sealants applied to P91 steel substrate coupons.[6] Corrosion tests were performed in the lab as well as in the field. A WTE environment was simulated: KCl-K_2SO_4 eutectic mixture was applied (to simulate low melting point deposits) on the coated samples and representative test temperatures were used (525°C, 625°C, 725°C) in a HCl-containing environment. Alloy 625 coatings gave the best results with little, or no chlorine/oxygen being detected at the coating/substrate interface. The authors postulated that above the eutectic melting point (684°C), the dominant corrosion mechanism was molten salt corrosion, wherein fluxing of the protective oxides by the molten salt occurred. Below the eutectic melting point, general gaseous corrosion occurred. HVOF coatings and sealants provided enhanced corrosion protection.

A specific material improvement study evaluated the use of a suite of "acidic" ceramic coatings on burner tips of small WTE boilers – B_2O_3, TiO_2, SiO_2 and Al-phosphate.[8] The burner material was 253 MA SS, which is a form of 308 SS. It was found that the Al-phosphate coating provided the best sacrificial protection, reacting with the alkali environment in the boiler and enabling a stable oxide scale to form.

Over the past 5 years, researchers in Canada have explored various thermochemical techniques to convert raw biomass feedstocks into valuable streams at elevated temperatures. Corrosion testing of three alloys (316 SS, Duplex 2101 SS and Alloy 625) was done in hot dilute H_2SO_4 solutions at 150°C to simulate acidic pre-hydrolysis biorefining conditions. As with the previous studies, Alloy 625 performed the best in short-term and long-term testing. The authors postulated that Mo passivation led to the formation of stable, protective oxide layers.[9] Similar results were obtained in static autoclave testing using hot pressurized water at 310°C and 10 MPa to simulate hydrothermal liquefaction. Alloy 625 out-performed 316 SS and 310 SS. Addition of K_2CO_3 led to the formation of soluble hydroxides and increased corrosion rates.[10] Supplemental work done by the same group to understand the effect of organic acids on corrosion of Alloy 800HT and Alloy 825 found that addition of acetic acid slightly lowered the corrosion rates, likely due to change of solution pH to near neutral values, thereby decreasing the oxide dissolution rates.[11,12] The authors reported the existence of a double-layer oxide structure: an inner Cr-enriched oxide layer and an outer Ni-Cr-Fe mixed oxide layer.

A different group in Canada published a study last year where it was reported that biofuel feedstock storage corrosion is a bigger problem than corrosion of finished products storage tanks.[13] The raw feedstock ranges from standard vegetable oils to agricultural/animal waste, leading to unknown process conditions and corrosion issues. Hot rolled carbon steel was coated with six types of epoxy linings and tested against six feedstocks with TAN numbers ranging from 3 to 30 at 160°F-250°F. Satisfactory results were obtained in lab testing as well as in a mothballed refinery that had been repurposed to produce biofuels.

From a real-time corrosion monitoring perspective, most electrochemical instrumentation and probes cannot withstand the high temperatures of waste-to-energy applications.[14] However, a tool has been developed using the principles of electrochemical noise and utilized successfully in evaluating high temperature chloride attack on radiant and superheater sections of a WTE plant.

4. Conclusions

Converting municipal waste into energy is an important step in achieving a carbon neutral world but presents enormous mechanical integrity challenges.

1. Two primary corrosion mechanisms affect WTE systems: alkali chloride corrosion and molten phase corrosion.

2. High Cr-Ni-Mo alloys work best in WTE boilers and tubing; Alloy 625 has been documented to provide the best performance.

3. Alloy 625 also works well as a weld overlay and when applied as a HVOF TSA coating. The high temperature corrosion protection of such WTE systems was at par with those constructed using wrought 625 materials.

5. References

1. W. Adlhoch *et al.*, "High-Temperature Winkler Gasification of Municipal Solid Waste", Gasification Technologies Conference, 2000.
2. Y-C Seo *et al.*, "Gasification of Municipal Solid Waste", Chapter 7, Gasification for Low-Grade Feedstock, pages 115-141, 2018.
3. K. Salmenoja and K. Makela, "Prevention of Superheater Corrosion in the Combustion of Biofuels", Paper No. 00238, NACE Conference, 2000.
4. Y. Kawahara *et al.*, "Demonstration Test of New Corrosion-Resistant Superheater Tubings in a High-Efficiency Waste-To-Energy Plant", Paper No. 00265, NACE Conference, 2000.
5. W.B.A. Sharp *et al.*, "Superheater Corrosion Produced by Biomass Fuels", Paper No. 1308, NACE Conference, 2012.
6. L. Paul *et al.*, "Cost Effective Weld Overlay Alloys for Waste-To-Energy Systems", Paper No. 10199, NACE Conference, 2010.
7. D. Harvey *et al.*, "Mitigation of Biomass-Induced Corrosion of Superheater Tubes using Thermal Spray Coatings", Paper No. 2625, NACE Conference, 2013.
8. R. Pompe *et al.*, "Evaluation of Ceramic Sacrificial Coatings for Improved Alkali-Induced Corrosion Protection in Biofuel-Fired Boilers", Paper No. 2159, NACE Conference, 2013.
9. M. Liu *et al.*, "Corrosion under Hot Dilute Acidic Pre-Hydrolysis Biorefining Process", Paper No. 15438, NACE Conference, 2020.
10. M. Liu *et al.*, "Corrosion under Hydrothermal Liquefaction Biorefining Process", Paper No. 15439, NACE Conference, 2020.
11. M. Liu *et al.*, "Influence of Organic Acids Released During Hydrothermal Liquefaction of Biomass on Corrosion of Candidate Reactor Alloys", Paper No. 16612, AMPP Conference, 2021.
12. M. Liu *et al.*, "Corrosion of Constructional Alloys in Corrodents-Containing Catalytic Hydrothermal Liquefaction Biorefining Conditions", Paper No. 18940, AMPP Conference, 2023.
13. R. Nagarajan *et al.*, "Biofuel Feedstock and Finish Products – Linings Case Study", Paper No. 18898, AMPP Conference, 2023.
14. W.M. Cox *et al.*, "Real Time Monitoring of Corrosion and Fouling Cost-Effective Weld Overlay Alloys for Waste-To-Energy Systems", Paper No. 10199, NACE Conference, 2010.

172c Life-Cycle Analysis of Recycling of Post-Use Plastic (PUP) with the Greet Model

Pahola Thathiana Benavides, Ulises Gracida Alvarez, Troy Hawkins, Michael Wang

In the United States, less than 9% of post-use plastic (PUP) is recycled which causes the loss of a valuable resource with high energy and carbon content. Recycling technologies, such as mechanical, pyrolysis, gasification, and enzymatic hydrolysis, enable the conversion of PUP to valuable intermediate chemicals and develop markets for recycled feedstocks, promote a circular economy for plastics, and reduce the dependence on fossil resources. Life-cycle analysis (LCA) measures the environmental benefits of these recycling technologies. In this presentation we showcase different LCAs of PUP developed in the Greenhouse gases, Regulated Emissions, and Energy use in Technologies (GREET) model that evaluates the greenhouse gas (GHG) emissions, fossil energy, water consumption, and solid waste impacts of converting PUP into new high-density and low-density polyethylene (HDPE, and LDPE, respectively) and polyethylene terephthalate (PET) plastics or valuable products such as lubricants. Topics such as the mass balance approach, co-feeding of recycled and conventional feedstocks, material flow analysis will be included in the presentation. The methodologies and approaches discussed in this presentation could be expanded to other products and technologies.

AUTHOR INDEX

Abousrafa, Aya 18
Adewunmi, Ahmad 4
Akin, Oguzhan 29
Al-Harthi, Mamdouh A. 21
Alotaibi, Mohammad 3, 20
Al-Rawashdeh, Mamoun 18
Alvarez, Ulises Gracida 75
Bautista, Lucylitzy 44
Benavides, Pahola Thathiana 75
Bhagat, Ashish 5, 6
Davis, Russ 67
Doyle, Stephen 22
Fahmi, Mohanad 3, 20
Geem, Kevin Van 29
Gonzalez, Ana 44
Goulas, Konstantinos 42
Guma, Erdin 65
Han, Minkyu 43
Hawkins, Troy 75
He, Qing 29
Herzog, Charles 48
Hoekstra, Hsiang Yee 66
Hussain, Syed Muhammad Shakil 3, 4, 20
III, Eloy Flores 66
IV, James Beall 7, 17
Janvier, Ben 17
Kamal, Muhammad Shahzad 3, 4, 20
Katebah, Mary 18
Khanlari, Ana 2
Kshetrimayum, Krishnadash S. 43
Lee, Younghee 43
Linke, Patrick 18
Lodha, Sanjay 24
Mahboob, Ahmad 3, 20
McMillan, Greg 8
Moosazadeh, Mohammad 44
Murtaza, Mobeen 3, 20
Patil, Shirish 3, 4, 20
Phd, ... 2
Romero, Julia Jimenez 22
Shantz, Daniel 25
Shehzad, Farrukh 21
Stubbs, Randy 23
Sutton, Ian 45
Svadlenak, Scott 42
Swain, Douglas 23
Taghikhani, Vahid 44
Trivella, Filippo 47
Varghese, Robin John 29
Venkat, Veshal 66
Wahid, Safiyyah 1
Wang, Michael 75
Wang, Yihan 29
Yang, Zekun 22
Yazdani, Parviz 29
Zhang, Nan 22
Zhu, Chenguang 22

AIChE
120 Wall Street, FL 23
New York, NY 10005-4020

ISBN 979-8-3313-1715-7

9 798331 317157

CURRAN ASSOCIATES INC.
proceedings.com